MY LIFE AMONG THE UNDERDOGS

MY LIFE AMONG THE UNDERDOGS

A Memoir

TIA TORRES

wm

WILLIAM MORROW
An Imprint of HarperCollins*Publishers*

HarperCollins books may be purchased for educational, business, or sales promotional use. For information, please email the Special Markets Department at SPsales@harpercollins.com.

FIRST EDITION

Designed by Bonni Leon-Berman

Library of Congress Cataloging-in-Publication Data has been applied for.

ISBN 978-0-06-241979-8

19 20 21 22 23   RS/LSC   10 9 8 7 6 5 4 3 2 1

Thinking back to how VRC first came to be and to where we are now . . . who would've thought? Certainly not me. From the wolves to the Pit Bulls and now the parolees, we as the underdogs have tackled some pretty big hurdles and fought some even bigger battles.

But as probably the most controversial nonprofit organization around today, we couldn't have done it without your help. And by that, I mean without you—the fans, our supporters, and viewers—Villalobos Rescue Center simply . . . would not exist.

This book is dedicated to you.

CONTENTS

MY LIFE AMONG THE UNDERDOGS

INTRODUCTION

Coming into this world a descendant of the most famous werewolf in history can have its perks—but as legends have it, it can also be a curse. My future was set from the second some priest tried to drown me in a pool of holy water. I kicked and screamed bloody murder: This was no way for a *loup garou* to begin her life. From that moment on, I was one angry little pup.

Like Lon Chaney Jr.'s Wolf Man, my exterior persona was not always on point. No matter how much I tried to toe the line, that damn full moon kicked my ass every time. So I guess it's no mystery why I was drawn to certain animals from an early age and have lived my life surrounded by them: wolves, black cats, and Pit Bulls. I have spent my entire life trying to defend the underdog—which usually involves fighting some of the cruelest humans imaginable. They continue to throw me to the wolves, only to find that I've returned as the leader of the pack. Like the alpha bitch I am, I am protective of my young, loyal to those who deserve it, and willing to take a silver bullet straight to the heart for what I believe in. A true pack leader may seem to stand apart, but she is never alone—and that is certainly true of me. Today, I draw my strength from my pack at Villalobos Rescue Center, my human family and colleagues as well as the canines. I suppose you could say they are my full moon:

They make me stronger, more intuitive—and much, much crazier.

★ ★ ★

As a child growing up in a functioning dysfunctional setting, it was the animals I was constantly surrounded by that kept me from jumping off that cliché of a cliff. Animals were not only my companions but my heroes, my friends, and my reason for getting up in the morning. The hundreds of hours I spent on the backs of horses ensured that I'd always be a free spirit—but my relationship with dogs didn't really start until I left the house at the rebellious age of seventeen.

Truly on my own for the first time and in need of some bare necessities, I took my last few dollars to Kmart (which, for you youngsters, was our Walmart at the time). In the parking lot, I encountered some people giving away puppies out of a shopping cart. Most of the black-, gray-, and white-speckled pups were quite flashy, but the one that caught my eye was the runt cowering at the back of the cart. As I reached in to pet her and make silly little "coo-coo" noises, she jumped up and nipped me right on the nose. Then she sat back and gave me the stink-eye.

I had to have her.

Cougar and I became inseparable. For years we traveled the rodeo circuit, living out of my truck when we had nowhere else to go. The bond between us was something new for me, something that couldn't be measured. Although small in stature, she had the heart of a mountain lion (thus her name), and when she got pissed at someone— or just felt wary of them—she would let out a slow, high-pitched siren of a warning. Then, lowering her head and fixing her deep brown eyes into a laser glare, she'd back that enemy into a corner like any wild creature you might see on a *National Geographic* special. Cougar was a badass, plain and simple. We could've been twins, if I were a dog or she were a human.

Cougar showed how mighty she was on numerous occasions, but never more bravely than when she sensed I was in an abusive relationship. Putting herself between me and whatever jerk thought I made a good human punching bag, Cougar would snap and lung at my attacker until I could get away. I learned a lot from that tough little cookie, who not only saved my life but served as a better role model than most of the people I knew back then.

In the summer of 1982, at age twenty-two, I was at a crossroads. Feeling that my life was going nowhere and that I needed a change, I enlisted in the army. As I prepared for eight weeks at boot camp, I promised Cougar I would be back for her, with a good career under my belt and enough money to take care of us. I left her in the care of my father and his new wife.

Life at Fort Dix, New Jersey, confirmed for me that I'm just not an East Coast kind of girl. I hated the humidity (which is ironic, since I now live in Louisiana) and found life there just too fast paced. I counted the days until I could return to California and start a new chapter with Cougar. We were confirmed "road dogs," and I couldn't wait to fire up my rust-bucket '65 Chevy truck and head out into the sunset, like in an old country music video.

Nothing could've prepared me for the heartbreak I experienced when I arrived at my dad's condo. Cougar had become "a problem," he told me. They'd gotten rid of her. Not much has ever moved me to tears, but I broke down sobbing right there and then, begging and pleading with him to tell me where they had taken her. I intended to go after her and beg her new owners to give her back to me. I'd explain what had happened. I knew they'd understand and feel terrible and let Cougar go. She'd run out their front door and jump up into the bed of my truck and life would go back to normal.

Without emotion, remorse, or even a decent apology, my father's Stepford Wife put her expensive wineglass up to her lips, took a long sip, and told me that they'd taken Cougar to an animal shelter.

Cougar was dead, and I knew it. Because of her behavioral issues, there was no way that any shelter (especially back in the eighties) would've adopted her out.

And just like that, all that I'd gained in the army—my skills, my uniform, my sharpshooter's pin, my military ID that said I was somebody special—none of it meant anything to me. The guilt I felt over that dog was overwhelming. I'd trusted my father, someone who should have been worthy of that trust, and he had betrayed me. To this day, I have never forgiven him, nor have I spoken to him. I don't even think about him. But Cougar is another matter. I think about her all the time. She lives in every dog I meet. She was my inspiration for the deep and abiding love of dogs that has shaped my life and purpose ever since.

★ ★ ★

After that trauma, I was dogless for around thirteen years. I went through a lot of changes in that time. I was a waitress, a country-western singer, I competed in country-western dance competitions, you name it. If it could make me a little money, I tried it.

Then, out of absolutely nowhere, my life became one long, out-of-control, off-the-tracks roller coaster ride, with dogs in every seat but mine. My estranged brother (who came around only when he needed something) showed up at my door with a huge white fluffy wolf hybrid, whom he called Cujo. He looked more like an oversize white teddy bear than the fictitious killer dog he was named after. The bottom line was: I had a dog again, just when I didn't need one.

I already had my hands full with two "cubs" of my own: my daughters, Tania and Mariah, who were still very young at the time. But after my experience with Cougar, I sure as hell wasn't going to let my brother leave with that dog. Without saying a single word, I took the leash from his hand and walked my new best friend inside, slamming the door in my brother's face. That was the last time I saw or spoke to him.

I knew nothing about these wolf/dog mixes that had become popular since the release of Disney's *White Fang,* but I connected with them on a deep level. I understood that they were caught between two worlds, unwanted in either one. The people who adopted them soon grew disenchanted with their "ways of the wild." The wolf rescues didn't want them because they were part dog, and the dog rescues were hesitant to take them because they were part wolf.

I was hooked. I began to learn everything I could about them. Wolves soon occupied my every waking thought and invaded my dreams. The more I read about them and studied their behavior, the more I was drawn to their enchanting spirit. The more time I spent around them, the more I became one of them. It felt as though my senses of sight, sound, and smell had heightened. Was I becoming more like the wolves—who are hyperaware of their surroundings and highly suspicious of strangers—or had I been like them all along? It hardly mattered. I felt myself slipping into their world and carrying my kids with me. We became true pack members.

By 1993, I was running a full-blown rescue for wolves and wolf/dog mixes. On a high mountain in Agua Dulce, California, Villalobos Rescue Center (Spanish for "Village of Wolves") was fast becoming a refuge for both animals and people who felt they didn't belong anywhere else. And for the first time in a very long time, I felt connected to something extraordinary.

It didn't take long for me to make friends within the rescue community. My skill at handling wolves and wolf hybrids prompted people to call me whenever an animal proved more challenging than the average Fido. That's when Tatanka muscled her way into the picture—the first of many beloved Pit Bulls to come. Our adventures together in Sri Lanka on the set of *The Jungle Book* inspired one of my favorite stories in this book.

Wolves and their kin were still the main focus of Villalobos, but it wasn't long before the little muscle dogs were competing for kennel space, and it did make a kind of sense. The two "breeds" had a lot in common, if not physically then socially. Both were misunderstood and maligned as vicious, in news stories and storybooks. It wasn't the animals that needed rehabilitating—it was their image. Was it even possible to chip away at centuries of bad PR?

Looking back, I can't help wondering what I was thinking—but I guess the answer is that I *wasn't* thinking. Not with my head, anyway. It was all part of my wiring. The best way to get me to do something is to tell me I can't, especially if I know in my gut it's the right thing to do. You'll find that most of the stories in this book are about animals (and a few humans) who needed nothing more than someone to believe in them and a purpose in order to show their true nobility. As tough as my road has been, I'm proud to have fought and advocated for animals whom others were content to exploit and destroy.

* * *

For years, I pushed forward, taking in all the animals I could handle and involving myself in the politics of the shelter world, which was slowly coming to its senses about Pit Bulls. I spoke often to city councils and such—sometimes assisted by my tiny daughter Mariah and her devoted companion L.A. Those two changed more hearts and minds in ten minutes than I could've done in a lifetime.

* * *

We soon became the largest Pit Bull rescue in the country and maybe even the world, but life in the desert was becoming impossible. The hurricane-force Santa Ana winds were wreaking havoc on a nightly basis, ripping off roofs, destroying kennels, and setting off wildfires all around us. On top of that, the Golden State's economy was going down the toilet. Donations had pretty much dried up, even as our dog population grew unabated. It was starting to feel like a war, with me and my kids on one side and Mother Nature on the other—and she was winning.

A chance visit by a relative of one of our volunteers set events in motion that would lead to the next chapter in the saga of Villalobos. The volunteer's brother had just been released from prison and she didn't want to leave him alone to "get himself into trouble." No problem. We declared it Bring Your Parolee to Work Day and welcomed our newest volunteer. Mark ended up staying for quite a long time, the first of many human underdogs within our gates.

In case you thought the idea of hiring parolees to work with our dogs was a made-for-TV gimmick, I promise you, that couldn't be further from the truth. The formerly incarcerated soon became a part of the fabric of Villalobos, and something about that fact prompted *LA Weekly* to write about us and name me one of the city's "Most Important People."

Honestly? The article came just in the nick of time. When it ran, every bank account we had was overdrawn, our vehicles had zero gas in them, and we were reduced to begging for food from the local church (not easy for my band of nonbelievers). I was seriously contemplating closing the rescue, all the while tossing and turning each night thinking about what might happen to our dogs.

That article and that award quite literally saved a lot of lives. Television production companies began to approach us about participating in a reality show. My first reaction was "Get lost!" (Or saltier words to that effect.) The wolf in me will not be denied, and I have always been a private and guarded person. I held out for about two weeks, before the grumbling in my belly and the fact that my kids were reduced to stealing toilet paper from a local fast-food restaurant brought me to my senses. Or what was left of them.

Some of the producers who contacted us were hoping for "parolees fighting" and "dogs raging." Needless to say, that was a no-go. Others, from what I could tell, would turn our lives into the punch line of a bad joke. I was beginning to think the whole thing was a bad idea when a new contender came forward, a production company by the name of 44 Blue. They had done the shows *Cell Dogs* and *Lockup*. Somehow, they had earned the respect of inmates in prisons all over the world, so I figured they would "get us" and present us in the right light.

Next came the search for a network. It would take another year or so of starving and struggling and hanging on to the dream before we'd get the news for which we had been waiting. We not only had a home, we had an entire planet: Animal Planet.

As happy and relieved as we were, I went into the process with a sense of dread. Opening up my life and home and workplace to a television crew day in and day out would be a big adjustment for anyone; for me it sounded like hell on Earth. (Even writing this book has been a huge struggle for me, and it's just words on paper.) But I was determined to do it, for the sake of the dogs. But I was adamant about one thing: Everything they filmed had to be authentic. There would be no scripts or "faking" of scenes. This reality show would be the real deal or no deal at all.

So . . . I was all in. My family was another matter. I'd been dragging them into everything I'd ever done their whole lives, and, mainly, it had worked out to everyone's satisfaction. But that didn't mean they wanted to be TV stars. What the hell had I brought down on their heads now?

None of my kids were jumping for joy about being on

television—and the truth is, I was proud of them for it—but they understood the importance of the opportunity and reluctantly agreed to participate. They shared my own priority: to keep the rescue open at all costs. I guess I hadn't done such a bad job of parenting after all.

As I hoped it would, the show struck a chord in viewers—a fact that I chalked up to my insistence that everything be absolutely real. *Pit Bulls & Parolees* was an immediate success, much to the delight of the network. But if you think it made us rich, think again. Even after a couple of seasons of steady growth in viewership, our finances were stretched to the breaking point. The cost of living in California was way out of control.

After an attempt to build a new rescue (and one that was much more affordable to maintain) about an hour north ended in a devastating fire (so much for the luck of the Irish), I knew California was over for us. We had to expand our horizons. I went to the local AAA office and picked up maps of several states that sounded both interesting and affordable. It was instantly obvious that we couldn't afford the East Coast any more than the West; I was sick of the desert; and I definitely didn't want to go anywhere cold. What I did want was green grass, lots of trees, and warm weather most of the year. That narrowed things down to . . . the South!

From that point, we let Animal Planet help make the decision. Their thinking was this: Wouldn't it make great TV to move to New Orleans, where that bitch Hurricane Katrina had created an epic problem of stray dogs?

We talked it over. While New Orleans sounded like fun, my family of desert rats had never lived in a city before. Our home had always been at the end of a long dirt road to nowhere. It would be an adjustment, for sure, but we all voted yes, and the network agreed to finance our move. *Pit Bulls & Parolees* was coming to the Big Easy!

Our trips across the country with the dogs and belongings were intense, backbreaking, and sometimes downright comical. I arrived with the last load of dogs on January 1, 2012. If ever there was a labor of love, putting the New Orleans place together was one. And making it even more special were the people of Louisiana.

Never had we felt so welcome and so appreciated. Never had we experienced such kindness from neighbors. But getting used to city life—and not just any city but New Orleans—proved as challenging as I'd feared! As a confirmed country mouse, I was more than a little bit out of my element. And as a public figure with a growing fan base, I couldn't just fade into the woodwork. Things became downright chaotic, and fast.

Even as we round out our tenth season of the show, I'm still in denial about being a public figure. I block out the stares and whispers and sneaky cell phone picture taking, and I try to go about my business as if none of it is happening. I know that real celebrities deal with this stuff differently, but I can only be Tia—no other options have ever presented themselves. And when the day comes that we have filmed our final episode, well . . . let's just say I'll have mixed feelings about it. At best.

But that doesn't mean I'm not grateful for all that *Pit Bulls & Parolees* has done for all of us. Truly, the impact the show has had on people's perceptions, and on the fates of thousands of animals, is beyond anything I could have imagined. The fact that some brave television people wanted to take a chance on a show that isn't exactly *Lassie*— and that then audiences embraced it—is heartwarming to say the least.

But when you do animal rescue work on a large scale, no amount of success is ever enough. There are always more mouths to feed, and the story has to go on. It is our fans, supporters, and donors who keep us all alive, and the TV show has been our way of reaching out to them.

You could say that my life has been a series of rescues in another sense. If the TV show hadn't come along when it did, Villalobos Rescue Center would not be around today. If we hadn't moved to Louisiana, we wouldn't have experienced the renaissance we needed as an organization, or met the great people and animals who keep us going. I wouldn't have had my heart and soul named Lucky, or Jethro, who makes me scream . . . and laugh. My kids wouldn't have met their soulmates. We came to New Orleans in the wake of a terrible tragedy, but so much good has come from all that bad, and I can truly say that we've found our home here, just as I try to find "forever homes" for my dogs.

Okay, so Tia Torres is not always the best "people person." Sometimes I fight when I should give in, and sometimes (not often) I give in when I should fight. It can be downright exhausting being Tia.

But thankfully, there were those who saw the good in what we do. We managed to take a show based around a certain type of person and a specific genre of dog and create an empire that so many can stand behind. Whether you're a cat lover, sleep with Chihuahuas, or don't own any companion animals, there seems to be something with our stories that draws these others in. To have police officers and judges, or even someone's grandmother, stop to shake our hand, well . . . that's the ultimate pay-off.

What I'm trying to say—and I really mean this—is that our fans and supporters are the stars of the show as far as I'm concerned. One of the reasons I wrote this book is to humbly offer my thanks. Also, to celebrate and honor a number of the very special dogs who have come into my life. I think you are going to enjoy meeting them.

While I'm in a thanking mood, I want to bow to the wolves who set this story in motion. To my extended pack members: It is your strength, resiliency, and honesty that taught me how to survive and live honorably. You taught me how to raise my human cubs with dignity and respect. As long as I'm around, your legacy will not go unrecognized or uncelebrated. On each full moon, I howl in your honor. You were the original "Pit Bull," and I will forever be a member of your pack.

Now meet some of the greatest dogs ever to grace Planet Earth.

1

TATANKA

Buffalo Girl

I was raised by wolves . . . literally. Not in the "running naked through the trees" sense, but I did actually live among them. And just like in the movies, over the course of many full moons I slowly began to feel like one of them. My senses became heightened. I understood them and I felt that they understood me. We were family.

That is as close to an explanation as I have for why, later in my life, I developed a desire to protect and save them. And it explains the name of the nonprofit wolf and wolf-hybrid rescue facility I would establish in the nineties, in the high desert hills of southern California: Villalobos Rescue Center.

But I'm getting ahead of myself.

I was raised by a single parent—my stepmother—and despite our struggles together, the one thing that kept us going was our love for animals. My "mom" instilled in me the importance of taking on the responsibility of caring for animals of all kinds, and she taught me that it was a privilege to share our lives with them. I grew up with mainly horses, but it was this undeniable draw toward dogs—actually canines of all kinds—that kept pulling me in.

While most teenage girls plastered their bedroom walls with pages from *Tiger Beat* magazine, mine were a scrapbook of dogs, coyotes, and wolves. It was as if I wanted to be one of them; to run and hunt with their skill was something I dreamed of at night. And while most young girls my age were begging their parents to get their ears pierced, I wanted a set of piercing yellow eyes, just like the wolves of Alaska had.

★ ★ ★

Many years and a few "skin babies" (that's what us animal people call our human babies) later, I would get my wish—well, sort of—when, in the early nineties, my estranged brother showed up on my doorstep with Cujo, his "failed" (that is, not sufficiently badass) guard dog.

Cujo was the most magnificent creature I had ever laid eyes on—and nothing like his bloodthirsty fictional namesake. He was a white avalanche of a wolf/Malamute cross, 120 pounds and 31 inches tall at the shoulders, and he stood before me as noble as an ice sculpture. I immediately noticed his eyes . . . the same yellow-amber eyes that I had dreamed of so many nights as a young girl. This dog belonged with me.

Despite getting dumped by my loser brother, Cujo showed no signs of distress or aggression, nothing that fit my expectations of how a "wolf" might act in the situation. He remained so calm and stoic that I was virtually hypnotized by his presence. All I could do was stroke his thick white coat in awe. And despite warnings from friends and family that this "baby-eating monster" would get a "taste for blood" and ravage my entire family, I felt there was something magical about him; that he was destined to do great things.

Although I wasn't aware of it, wolf/dog mixes were all the rage at the time. Thanks to the popularity of the Walt Disney movie *White Fang*, everyone wanted a little "piece of the wild" in the form of a pet—just as they had to have the latest fashions and accessories. Breeding and selling wolf hybrids was big business—never mind the fact that many of the animals sold as "part wolf" were really just overpriced Husky mixes. Many of the people who did end up with the real thing soon realized that they had bitten off more than they could chew. Between the howling, the destruction of property, and the menacing of neighborhood cats, these Disney creatures were more than a handful. Disgruntled owners started dumping them right and left.

Not needing a full moon to wreak havoc, Cujo and his pack quickly became the start of many more wolf hybrids to come. Sixty miles outside the City of Angels, my ten-acre desert oasis became the rescue site for these "dogs" that were tragically caught between two worlds. Some carried the behavioral traits of pure wolves, while others acted just like a typical Fido. But the very clear message here was that this was meant to be. I had finally found a place where I felt I belonged.

Right around this time, a distant family member revealed

a secret that would knock everything into perspective. Over lunch, I was told by a relative that apparently, I was the goddaughter of the actor Lon Chaney Jr.—the original Wolf Man. Yes, the very one who had battled Bela Lugosi, aka Dracula, in those old black-and-white movies. Now it all made sense and answered a lot of my questions as to why I'd always felt "different." I guess you could say it was in my blood—figuratively speaking, of course. And from that moment forward, I knew what my life's mission would become.

By the midnineties, Villalobos Rescue Center was in full swing, taking care of sixty wolves and wolf hybrids. Running it was really my third job, in addition to working at a pet supply shop and taking care of my kids. I started networking with other animal rescue people and was getting to know my way around that world when one of them, a Collie rescuer named Vickie, unwittingly helped push Villalobos in a new and extraordinary direction.

Vickie called me one day, hoping I would go with her to a local animal shelter to bail out a Collie. My first thought was, *Who the hell dumps Lassie in a shelter?* Back then I was still a little naïve about the number of beautiful purebred dogs that end up in the pound for one crappy reason or another. Still trying to learn my way around the shelter system, I agreed to join Vickie to learn the ropes.

I grabbed my daughters—Tania, then ten years old, and Mariah, four—and we all drove up to the remote high-desert town of Lancaster, California. It was a flat and barren place, filled with busted-up trailer parks and populated by drug dealers and users. Animal neglect went with the territory and, sadly, so did dogfighting, which was a huge problem in Lancaster. Vickie tried to prepare me for the worst, saying that I'd better grow some thick skin before walking down the rows of cages at the shelter. As we entered, I noticed a little grassy spot with benches and tables—the "meet and greet" area where potential adopters could get acquainted with the dogs. Rather than take my kids over to the cages of raging animals, I parked them on a bench while I walked over to check out some of the desperate souls.

As I tried to give some comfort to a frantic Labrador (while intermittently looking back at my daughters, who were behaving like the good little girls they were, giggling and tickling each other), I noticed a side gate open to a very disturbing sight. A sheriff's car and an animal control truck were parked, and several officers were walking a brindle pit bull in as if she were Hannibal Lecter. She was only about fifty pounds and, from my vantage point, it looked as if she didn't have any ears at all: They'd been cut off at her skull.

Then the unthinkable happened.

As the little nub-eared dog and her entourage came through the big chainlink gate, she slipped out of her rope leash and began to run toward my daughters. In between pissing myself and letting out a scream, my maternal reflexes kicked in. I catapulted from my spot and sprinted across the grass. The girls, oblivious to the charging animal, continued playing pattycake games.

Because of my work with wolves, I understood that I didn't need to be worried about the "big mean Pit Bull." This was simply: *Holy shit, a dog of unknown origin is going after my kids!*

Within a few steps, it was clear that I wouldn't be able to outrun the little athlete; she got to Tania and Mariah within a few split seconds and leapt up on them, all four feet leaving the ground. The bench flipped over backward. My heart dropped into my stomach. Everything unfolded in slow motion. An overwhelming wave of helplessness washed over me. And then I heard an unbelievable sound.

Giggles. Waves of little-girl giggles.

My daughters were on their backs with their legs sticking straight up in the air, and all I could see was this tiger-striped dog butt, its tail wagging so fast it was a blur. The "monster" was kissing them frantically, going back and forth between Mariah's face and Tania's ears. Never in all my years had I seen a dog love a couple of kids as vigorously as this muscle butt was doing.

By that point, the animal control officers and shelter staff had come running and everyone was making a fuss over us, making sure the girls were okay and apologizing over and over. Word spread throughout the shelter that there had been an "incident," and a crowd began to gather. I was stunned as I watched the officers pull the little brindle off of my daughters, who were actually disappointed that the "funny dog" was being taken away. Although my heart was still racing, I noticed the dog look back at my girls as she was being trotted off to her concrete prison cell among dozens of hysterical animals. Those few seconds of happiness would have to last her awhile, and I couldn't help feeling sad for her.

As we all left the shelter with the Collie we'd come to liberate, I couldn't stop thinking about what had happened. I had to know that little Pit's story. "Hold on a second," I said to Vickie—and I marched myself back inside to ask the clerk about her situation.

Turned out the brindle was on a "custody hold" for the sheriff's department, extricated from the scene of a double homicide in what turned out to be a meth lab. From what the officers could figure out, a drug deal had gone bad and everyone had started shooting. The only survivor was the little "guard dog," found chained to a truck. Her ears had been tied off with fishing line to cut the circulation so they'd fall off. I guess her owners thought that it made her look extra tough.

I was stunned at what this little dog had gone through. Yet, somehow, there remained a kind of innocence about her.

Days went by and I couldn't get this compact, exuberant dog out of my mind. What could I do? I called the shelter and asked if I could adopt her. Because she was considered evidence, they told me, it could take quite some time to release her. We'd have to wait until the investigation was complete. How sad and frustrating that an innocent bystander to a crime would have to sit in "doggie prison" until the authorities decided to let her go. It took about two months, but I finally got the call I had been waiting for, and back up to Antelope Valley I went.

★ ★ ★

My budding wolf/wolf hybrid facility sat atop a hill over-looking the ten acres that was now widely recognized as Villalobos Rescue Center. I found an empty kennel for my new resident and vowed to find her a new family deserving of the overwhelming amount of affection she had to offer. It pained me to click the latch shut and walk away from the grateful little dog, but she was a rescue and I needed to keep some distance between us by not bringing her into my home. I had my "house pack" already, and there just wasn't any more room. But that didn't mean she wasn't worthy of a great name.

The sweet little Pit Bull reminded me of something but I couldn't quite figure out what. Then it hit me. Although she was a "she," there was something masculine about her; she looked like a mini-buffalo! Since our lives revolved around wolves and wolf hybrids, it's not surprising that the tone of the place was a bit Native American. I remembered a scene in the movie *Dances with Wolves* where an Indian is trying to teach Kevin Costner's character the word for *buffalo*. He raises his hands up to the sides of his head and makes horns out of his fingers and says "*ta . . . tan . . . ka*" in that stereotypical Indian voice. From that moment on, the brindle was Tatanka.

As time went on, my little buffalo gal got sweeter—if

that was even possible. Although she was comfortable liv
ing among the wolves and wolf hybrids, I had to remind
myself that she was indeed a dog and that I needed to find
her a forever home. She'd had a horrible past and I knew
there might be some issues to work out first, so I decided to
bring her down to the house after all, for some training and
socialization. Although I'd never owned a Pit Bull, I knew
that they could be predisposed to animal aggressiveness
(damn that Terrier in them). I needed to see to what extent
that might be true of Tatanka.

For her first encounter with other animals, we stumbled
upon chickens. Now, don't get me wrong: I love animals of
all sorts, but having chickens had been the idea of my ex,
Jon. He loved them. He named them, and they followed
him around like puppies. He had this one nasty rooster
named Rex who would attack any one of us the second we
set foot outside the door. The kids would literally have to
carry a broom to fight off this beast. So, yes, a little part of
me was hoping that Tatanka would set Rex straight.

Well . . . despite the fact that I'm Irish, luck was not with
me on that day. Tatanka had nothing but love for these yard
birds. As soon as we let her out, she lay down on her side in
the sun, head stretched out. The chickens—including nasty
old Rex—nestled right down with her like something out
of one of those "cute overload" animal videos. My first reaction was relief and excitement that Tatanka was so good
with the birds—but this also meant that me and that damn
bird were going to have to work out our differences . . .
one way or another. More important to me, though, was

yet more evidence of how special this little dog was. Even chickens felt comfortable and safe around her. Tatanka was going to have a new home, all right—ours.

We moved her into the house permanently, and she promptly became fast friends with the other dogs and cats, breaking every stereotype of the breed. She bathed the cats; snuggled with the dogs, wolves, and wolf hybrids; and washed the horses' legs for hours at a time. From wild field mice to the range cattle, she loved all creatures.

But Tatanka's main love was children. When she spotted them, she would charge over to her next "victim," and the nonstop licking machine would go into high gear. That would inevitably trigger a giggle-fest so infectious that anyone who happened to be in the vicinity would collapse in helpless laughter.

Tatanka did have her rules, though. With older children, she began with the cheeks, then moved on to the ears—outer and inner. The laughter she invoked came from somewhere so deep that it sometimes sounded as if the children might puke. And the more they tried to wiggle away from her caresses, the harder she pressed herself into them, sometimes pinning them to the ground with her front feet on either side of their head so that she could continue her tongue bath.

But that was nothing compared to her affection for babies in strollers—whom she could spot from a mile away. The licking of squishy baby toes and food-stained baby cheeks was what Tatanka lived for. Where on earth did this love of children come from? She certainly didn't learn it during her horrific younger years. Maybe dreaming of a family of her own was how she got through each painful day chained up to that truck axle. Whatever the case, we had a true treasure on our hands.

★ ★ ★

Around this time, the reality of managing a wolf/wolf hybrid rescue operation was slapping me upside the head. As drawn as we were into the "wolf culture" scene, the forces of nature were working against us. Because I was dealing with canines that were neither wolf nor dog, it was difficult to find people willing to donate to a cause that catered to both. Wolf people love wolves and dog people love dogs . . . it's just the way it is. And, sadly, my "mutts" were a mix of both. With the lack of funding, the "Village of Wolves" was struggling.

I was up every morning at six o'clock. Situated as we were in the high desert, the weather was brutal nine months out of the year. And in those early "wolf years," I had only my good friend Nicole Wilde to help me. Together, we mucked out kennels in rain, snow, and 75 mph Santa Ana winds— then we ran off to our "real" jobs. For me, that meant stocking inventory, dealing with customers, and other menial tasks at the local pet supply store. On my lunch break, I'd race home to check on the beasts, then hurry back to work until the evening.

Throughout all of it—even at work—Tatanka was my constant companion. Although she didn't have a wolf's silver coat and her eyes lacked that piercing tundra-yellow stare, she had grown on me. Her days were now spent sprawled out behind the counter at the store, and when that front door went "ding-ding," she was up and trotting over to whomever walked in to offer up a friendly kiss and a muscle-butt wiggle. She was more than happy to escort customers up and down the aisles like the "Employee of the Month." So late one afternoon, when I heard an unfamiliar and uncomfortable bark out of Tatanka, I knew something wasn't right. This wasn't her "Can I help you pick out a new doggie toy?" kind of greeting. This one made my hackles go up.

Tatanka jumped up from her late-afternoon sunbath and ran to the back of store with her tail tucked between her legs. A second later, a man and a woman sauntered—or maybe staggered—in. They were typical of some of the types who lived on the fringes of our desert community, and there was no doubt that they were either drugged or drunk out of their minds. As the sole employee in the place, I felt my palms begin to sweat and my nerves jangle like the bell on the front door. It was then that Tatanka ran to the middle of the store and began barking frantically. I could see she was afraid but trying hard to be brave.

The woman yelled out, "Mary? Is that you?" And she tried to get a closer look at Tatanka.

I stepped in front of her and the stumbling man, trying hard to appear confident but not confrontational.

"Can I help you?"

I wanted them out of the store—which was located on

the outskirts of town, pretty far from any kind of help—and was trying hard to think of a Plan B if things started to go south.

The woman tried to shove me to the side in an effort to get to Tatanka, who continued to bark hysterically.

"That's my sister's dog! It's her! My sister was killed, and *that's her dog!*"

At this point, Tatanka gave up the bravery act, retreated into the office, and hid under the desk.

It took me a minute to figure out what was going on, but Tatanka's behavior told the entire story. She knew this woman and had been petrified from the second the couple walked into the store. I tried to appear casual as I locked Tatanka in the office and put on my best game face. By that point, the woman was quite irate but—thankfully—she could barely stand up.

I weighed my options: Should I be "typical Tia Torres" and let them know in no uncertain terms that they weren't getting their dog back? Or should I take the easy way out and say they had the wrong dog? I had to do what was best for poor Tatanka, who was still barking hysterically behind the locked door.

I folded. I did my best to convince the scuzzy couple that it was a case of mistaken identity. As I spoke to them as calmly as I could, they seemed to forget why they had even come in. I managed to ease them out the door and lock it, though it wasn't yet closing time.

I then coaxed the tough-looking yet very sensitive Pit Bull out. She peeked from behind the office door, and I convinced her it was okay. We both sat there on the floor and held each other. It was now up to me to make her feel safe again. She had seen the demons from her past, and I vowed to never let that happen again. At that moment, I made the decision to protect her at all costs, always. I never wanted her to leave my side, and right there, on the floor, in an embrace that felt like forever, it became Tia and Tatanka. Together we would change the world.

We became inseparable. Nothing could keep us apart for even a day. So, three years later, when I was approached by a movie company to work for them with my wolves, it was exciting news that I wanted to share with Tatanka! I had no experience in training animals for the film industry, but the company would train me in exchange for permission to use a pack of my wolves and wolf hybrids to retell the story of *The Second Jungle Book—Mowgli & Baloo*. Just the thought of such an opportunity was unbelievable—but I had a lot to consider.

I knew this would involve real sacrifice. I would be away for almost a year, training exotic animals from every branch of the animal kingdom. I would have to leave my kids with my "baby daddy," Jon, who'd been in my life for twelve years. Although he and I were no longer together, he was still a great dad. He agreed to take care of both girls so that I could take advantage of this once-in-a-lifetime chance.

But what about the rescue center? Nicole and Vickie

would have to look after everything. It was a lot to ask and I was torn, but after much discussion all around, it was decided that this was something I couldn't pass up. Jon would play Mr. Mom in my absence, and Tatanka would travel around the world with me. After months of prep, we loaded up what felt like Noah's Ark and prepared for our trip to the jungle.

The cargo plane was spectacular. It was divided in half. The back section was like a mini-warehouse area, and in it were various types of cargo stacked in big wooden crates and boxes. The front half of the plane would be filled with jovial yet unsuspecting passengers, most unaware of the exotic patrons who would share the journey with them in what could be best described as a "zoo with wings."

I watched as tigers, a lion, bears, monkeys, leopards, and the wolves were loaded up. Tatanka was the last to board, and she was not happy about it whatsoever. She had to be crated for traveling purposes, and among all of the animals, exotic and domesticated, she behaved the worst—barking, whining, and clawing at the gate of her kennel. I felt horrible. I knew we'd have a two-day layover in Amsterdam, but first there was the ten-hour flight from the United States to endure. Between Tatanka's anxiety and my overwhelming fear of flying . . . it was going to be a long trip.

I spent much of the flight lying on top of the cage belonging to Romeo, the huge African lion. This served the dual purpose of keeping the neighboring Tatanka calm and helping me avoid barfing up my breakfast. Tatanka sat drooling in her carrier, and I realized I was doing the same. Romeo watched impassively as I moaned from motion sickness and fear. Flying was definitely not for either of us.

About nine hours after we took off, a six-foot-tall, blond, blue-eyed flight attendant came in to tell us we were landing. My first thought was, *How dare you look so damn good?* I mean, there I was, sprawled out on top of a lion cage, my face stuck with dried spittle to a crumpled napkin and my hair falling out of a eighties-style scrunchy, while Holland's Next Top Model was cheerfully making announcements. Tatanka had finally settled down but woke up as Romeo let out a muffled "good morning" roar. Thankfully, we were only one barf bag away from landing in the beautiful city of Amsterdam.

I lay on my back to keep my stomach from doing flip-flops and locked eyes with Tatanka. Here we were, traveling the world together, and I wondered if she remembered those horrible people back in the desert. It made me think about how resilient dogs are. We humans dwell on each painful moment, while dogs just shake things off and go about their day. You have to admire them for that. Tatanka was my hero, but I still had no idea just how much I had to learn from her.

Amsterdam was breathtaking. The facility the animals were to be housed in was state of the art and incredibly impressive, but I decided not to leave Tatanka there. Instead, she'd come with me and the two trainers with whom I was traveling, Scott and Chad. The four of us waited outside the airport for transportation for over an hour. I really needed a Coca-Cola fix and hunted for a vending machine, only to find that Amsterdam was a Pepsi city—and that a can of the stuff was three dollars. As I stood there cussing out the machine, Tatanka looked up at me and wagged her tail slowly, as if to say, "Look, Pepsi or Coke, who cares? You're still my girl." So I gave in and downed the pricey sugar water despite my loyalty to the other beverage.

I noticed immediately that Amsterdam was a dog-friendly city. They were everywhere. So we decided to give up on the production company's mythical "private car" (that never arrived) and hop on a public bus. A few people smiled as Tatanka very calmly sat down on the floor of the very crowded vehicle. Most of the passengers didn't even notice. We were finally on our way to the hotel. All was good, and I was settling in to see the sights out my window when my eyes found something else on which to focus.

Two men dressed in military-style uniforms boarded the bus—policemen, I soon figured out. Based on the behavior of the other passengers and their lack of concern, I gathered that this was nothing out of the ordinary. The men hung on to the handrail with one hand and steadied the automatic weapons strapped over their shoulders with the other. I couldn't help but notice one of the officers looked a lot like the actor Dolph Lundgren (during his hunky He-Man movie role), and he was giving me the eye. Of course, I couldn't resist and I smiled at him, completely unaware that his attention for me actually had zip to do with . . . me.

"Peet bool?" Dolph asked in his thick Dutch accent.

I nodded with a proud, "Yes, she is," not realizing that my Adonis from the Netherlands knew not a word of English. As he began to spout a stream of "Angry Amsterdam," my smile began to feel pasted on, but I kept nodding as if I understood what he was saying. It wasn't until another passenger on the bus who spoke a little bit of English popped my Dutch dreams.

"Peet bools . . . no good here. You go to jail!" he said, pointing to Tatanka, who had now fallen asleep, completely unaware that she was the subject of a serious conversation and about to be the cause of my incarceration.

"Wait . . . what?! You, I mean . . . they . . . don't like them?" I stammered as my face began to feel hot. I didn't like where this was heading. As my translator held out his wrists and made the handcuff motion, the bus driver began barking out the next stop over the loudspeaker. The once nonchalant passengers had begun whispering and pointing at me.

"Holy shit . . . Pit Bulls are illegal here," I whimpered to Scott and Chad.

Dolph and his partner motioned for us to get off at the next stop, and I began to imagine what an Amsterdam jail might be like. This was turning into an episode of *Locked Up Abroad*. I began to practice my "but I'm an American" speech. Surely when we told them that we were with a film production company they would be impressed into releasing us. Maybe they'd even ask for our autographs! But something told me that their wooden shoes were about to kick me straight in my mouth.

So there we were, standing on the sidewalk of a city in which prostitution was legal and you could sit down in a coffee shop and order hashish from a menu . . . yet we were about to go to jail for having the "wrong" dog. For some twisted reason, Scott and Chad thought it was funny. I guess they could afford to laugh—they weren't the owners of a notorious Pit Bull, just a few innocent bears and monkeys. I pictured my forty-five-pound "beast" and me sporting orange jumpsuits and bending over for our cavity searches.

I'm not sure if it was the tears or my "we're in the movie business" defense, but eventually the Dutchmen communicated to us that we should just get in the cab they flagged down and stay in our hotel room until it was time to leave the city. I'd have to live out my foreign jail fantasy some other time. With a bruised ego and my tail between my legs, I forked out the forty-five dollars for the taxi ride and bid farewell to what could've been my future ex-husband.

Tatanka and I did as we were told. After two days of living on room-service pizza and three-dollar cans of Pepsi and watching bubblegum commercials that could've passed for softcore porn, we were ready to head out to our final destination: Sri Lanka. It would be smooth sailing from then on. I mean, how much more trouble could one little Pit Bull cause?

Apparently, a lot.

★ ★ ★

Eleven hours later, our flying zoo landed at a tiny airport on the island once known as Ceylon—and I could see why it was chosen as the setting for the movie. As we passed outdated airplanes and donkey-powered carts hauling chickens, I felt as if I'd entered a time warp. As Westerners and members of a movie company, we were treated like royalty. The natives came from all corners of the airstrip to assist us with unloading baggage, equipment, supplies, and of course the animals. This was going to be an experience of a lifetime.

I was beyond exhausted, and while the other trainers who'd flown out ahead of us tended the exotic animals, my attention was on Tatanka, who was anxious to get out of her crate. I liberated her and began to walk her around the airport as she happily stretched her legs, oblivious to the strangeness of her new surroundings. That was one of her best qualities: No matter what you threw at her, she took it in her waddling little stride. In spite of the developing chaos as the exotic animals were loaded into trucks, Tatanka strained at her leash in an attempt to meet everyone.

Considering she had to compete with a lion, three bears, two tigers, three leopards, several chimps, and a pack of wolves, there seemed to be way too much attention being paid to Tatanka. People began to surround the two of us, pointing and chattering in a language I couldn't begin to comprehend. As much as I wanted to think they were thrilled to meet her, I did begin to have flashbacks of Amsterdam and my new spinoff show, *Women Behind Bars*. But the locals seemed to be making funny gestures with their hands, putting them up to their heads and laughing. Could they be making fun of Tatanka's little earbuds?

Just then, Brian, the animal training coordinator, came over with a man holding a clipboard. "Tia, there is a slight problem," he said quietly. "They can't find your dog on the flight manifest."

"What? Of course she's on there. I watched the guy in the airport do her paperwork," I said. I walked over to her crate and took her records out of the little plastic pouch attached to it.

"No, you misunderstand," continued Brian. "The manifest shows a *dog* but . . . these people think she's a . . . baby bear!"

I searched his face to see if he was making a joke, but I could tell he wasn't kidding. Luckily, our interpreter showed up and explained the situation more fully. Apparently, most of the dogs in Sri Lanka are small and very mangy. There were strays everywhere, most of them covered in a contagious skin condition called sarcoptic mange, and they were not looked upon kindly, to say the least. Because Tatanka was relatively large and muscular, with a shiny brindle-colored coat and those little nubby ears . . . okay, I could see how they thought she resembled a bear cub. But surely, they didn't really think she *was* one!

Hours later—three, to be exact—we still had not succeeded in convincing the airport staff that Tatanka was a dog. Dogs just weren't that "pretty," they said. But a few American dollars later, placed into the palms of some airport staff, and our animal convoy was finally on its way.

We set up camp in a small village in the jungle surrounding the city of Kandy. Tatanka lived among the rest of the menagerie and served as a kind of calming agent for the other animals. She would saunter from cage to cage, greeting each with a swagger and that stereotypical Pit Bull wag. And like every Pit Bull I'd ever met, she grabbed a few human hearts as well as the animal kind.

Late every afternoon, Tatanka would sprawl on the wooden deck next to the enclosure belonging to Shadow, the black leopard, toes stretched out as far as they could go as Shadow's sleepy paw reached through the cage and patted her head. Then the two would drift off into an afternoon slumber together. As leopards go, Shadow was gentle—nothing like the sinister character Bagheera he was hired to play in the movie. I had no idea at that time that he and Tatanka would remain friends for years to come.

Spirit, the wolf, was not as welcoming a friend. Word around the compound was that she was an alpha bitch, a tough old broad with humans and animals alike. But somehow that "old soul" Tatanka managed to tame even Spirit. They spent many a day lounging together and watching the squawking parrots take flight and the mischievous rhesus monkeys chatter and chase one another around the high jungle canopy.

During the almost one year we spent living in Sri Lanka, my special pup met and befriended more varieties of animal and human being than you can imagine. She once encountered a group of Buddhist monks who thought she looked like a monkey. Cobbling together our broken English and

Sinhalese, we dubbed them "Chunky Monkey and the Monks."

The breed was basically unknown on the island, and to watch the unsuspecting locals greet her with smiles and compassion was refreshing. She was not being judged for what she was but rather *who*—and she left a mark on everyone's heart. The people of this somewhat Third World country were so kind in welcoming us into their homes. Whether entering a wooden shack in the middle of the jungle or a palace surrounded by beautiful rivers, Tatanka put on her best smile and wiggled her most determined tail wag.

It was sad to leave Sri Lanka, a place so magical that a Pit Bull could live freely without persecution. But home was home . . . so we headed back to the United States.

To this day, I consider my time in Sri Lanka with the animals to have been life altering. The people there live so simply and go without so many of the things we take for granted—yet in many ways, they seem happier and more at peace than anyone I know in the States. In fact, it took me quite a while to adjust to being back. I felt as if I didn't belong in the hard place in which I'd grown up, with a dog who loved all creatures unconditionally but was met with hateful stares and suspicion wherever she went.

Not that I let those feelings get the better of me for long. I knew I was meant to do big things—I just didn't realize how big at the time. Juggling my work as an animal trainer for the movies and running a growing rescue center, my confidence grew and my thirst for saving animals could not be quenched. And I came away from Sri Lanka knowing that I had to do what I could to rehabilitate the undeserved reputation of Pit Bulls. Tatanka and I started making personal appearances with that in mind; I knew that if anyone could change the world's outlook on the breed, she could.

In 2005, Tatanka lost her battle with cancer, unaware of her own greatness, the minds she changed, and the hearts she won. Without her, I would never have known what it feels like to bond with the greatest breed of dog in America. People call me a hero all the time, but the truth is that Tatanka—the buffalo, the bear, the chunky monkey—is the heroic one. Without her, there would be no Villalobos Rescue Center and certainly no *Pit Bulls & Parolees*. To her, I bow down with respect for one last ear washing.

2

A DOG NAMED L.A.

The Princess and the Pit

Having now been in the world of Pit Bull rescue for a short time (although it felt like years), I was beginning to realize the extent of abuse and neglect that went along with this breed. Being located in Los Angeles County, I saw it with all animals, but there was something about Pit Bulls that just brought it to the forefront. This was all before the Michael Vick case, so the abuse and torture that the breed endured had really yet to be exposed in the media as it has been since. Some of the abuse cases we dealt with made the dogfighting ones seem almost mild. The Pit Bull was fighting for its life on so many fronts that the memory of it as the "nanny dog" or "Petey from the *Little Rascals*" was quickly fading.

Even at animal shelters, Pit Bulls were the pariahs of the dog world, considered "too dangerous" to be put up for adoption, and any dog rescued from a dogfighting environment was deemed downright untouchable. No matter how friendly and trainable he or she might appear, they'd be euthanized on the spot, by order of local law enforcement.

By the time I had expanded the mission of Villalobos to include Pit Bulls, I had managed to build up a good working relationship with the animal shelters in and around the Greater Los Angeles area. My work with wolves and wolf hybrids had proved that I could handle myself in and around "difficult" breeds—but my colleagues didn't know the half of it. The wolves had taught me more than any book or seminar could ever communicate.

And then there was my work with the exotic animals. After working with tigers, bears, and various other members of the wild kingdom, it was difficult for me to compare the danger element of a fifty-pound dog to that of the tigers, bears and various other members of the wild kingdom with which I had worked. With that notch in my belt, I was permitted to take custody of dogs where others might have been denied. Villalobos Rescue Center had gained the reputation of being realistic as well as very responsible. We did not feel that every dog could be "saved," as there were many factors involved in determining a dog's adoptability. Based on this attitude alone, we stood apart from many other rescue groups.

For years, humans had portrayed wolves as demons of the night, with their salivating fangs and piercing yellow eyes. They hunted little red-caped girls, and, during every full moon, packs of them ravaged villages, ripping the flesh from the throats of innocent people. They "huffed and puffed" their way into every child's nightmare and devoured grandmothers.

This undeserved and exaggerated reputation got them

slaughtered almost to the point of extinction. Truly shy and demure animals, they suffered through years of persecution and were accused of every unexplainable animal attack or death, every missing child or mystery killing of a human. The truly sad part is that wolves are timid animals and they fear humans. Shy and suspicious, they rarely show themselves out in the open. And—let's face it—if we're being honest here, man is the only animal that hunts purely for sport.

Yet it all sounded uncomfortably familiar. There was a pathetic parallel between the two worlds in which I had now gotten myself deeply involved. The ignorant "now that they have the taste of blood" idea had become an all-too-common sound bite for news reporters. And there was "no doubt" that once a dog was involved in a dogfighting ring, he or she might as well be labeled an offspring of Cujo, "bred to kill" (an uneducated phrase I'd heard all too often).

My family and I lived among these dogs and we saw nothing even remotely close to this behavior. Either I was truly naïve, or society was that paralyzed by ignorant fear. I was rescuing the two most infamous canines in the world, yet to me, both were simply victims of a myth. I was up against a worldwide battle, and at times the frustration was unbearable. Many a night, I fell asleep with my head on my desk, hundreds of scribbled-on Post-it notes as a pillow. The countless phone messages describing Pit Bull abuse scenarios or owners giving up their "pet wolves" had taken over my life—what little it felt like I had left.

★ ★ ★

Funny thing about Los Angeles: It sounds like paradise to everyone who lives elsewhere. As someone who was born and raised in the "City of Angels," I can tell you that the reality is a bit more complicated. For every beautiful orange grove or stunning sunset over the ocean, there is an act of violence—and that includes violence against dogs. So when I got involved in my first dogfighting case, horrible as it was, I was prepared for it: pumped up and ready to kick some ass.

The call came from one of the many Los Angeles City shelters I dealt with, asking if I'd be willing to take some dogs who had been confiscated by the LAPD during a dog-fighting bust. Thankfully, there were only two of them, but one was severely injured. For some unexplainable reason, the first dogfighting case is seen as a badge of honor for a Pit Bull rescuer—not in the bragging sense, but more like "now you know what you're involved in." Back then I had no idea how many more fighting cases I would yet be a part of.

L.A. came "pre-named," somewhat unoriginally, for the city we both called home. When I first laid eyes on him, his tired body was blood soaked and his ears ripped to pieces. I could see patches of white hair here and there, but it was impossible to determine his true color.

During the drive home, every time I turned my head back to check on him, he'd give me a slow wag of the tail as our eyes met, as if to say, "You don't need to worry . . . I'll be okay."

He wasn't a big dog, so I was able to carry him from the car straight into the bathroom and place him in the tub. I turned on the faucet and warm water cascaded over him, beginning to provide some relief for his aching body. Once he'd settled in, I went out to the garage to get more towels. Just thinking about the pain he must be in made my heart ache.

Stack of towels in hand, I made my way back down the long hallway to the bathroom. *Hmm . . . must have left the radio on,* I thought, when I heard voices over the sound of running water. As I got closer, I recognized the unmistakably familiar sound of a giggling child: my child! I threw the door open to find a tub filled with way too much Mr. Bubble and four-year-old Mariah, my youngest, up to her neck in soapy water. She had scrubbed L.A. squeaky clean and both were wearing soap-bubble hats.

What was this child of mine thinking?! At four, I suppose there wasn't much thinking going on at all . . . and in her defense, she was my daughter. This is what my kids do.

I stood there frozen, a little hysterical but not wanting to make any moves that might upset either of them. My baby girl was frolicking in a tub full of bloody water with an abused dog we'd known for a matter of minutes. When I tell you that my concern had nothing to do with the breed, I truly mean it. My kids had been raised among wolves, after all. But a new dog with an unknown temperament is always a potential risk, and right about then I was trying not to panic.

In her quest to take care of the injured angel, Mariah had jumped into the tub with him and washed him as well as her tiny little hands could. I began to scold her but my words were drowned out by splashing and laughter as Mariah made soap-bubble sculptures on L.A.'s back. And, true to his kind, he was patient and gentle and adored every second of her attention. Those splashes were the result of his vigorous underwater tail wag.

The apple doesn't fall far from the tree, I thought. Mariah was definitely her mother's daughter.

The little girl and the little dog became inseparable, apart only when she went to school. And even then, L.A. accompanied us to drop-off each morning, walking Mariah all the way to her classroom and kissing her goodbye. At 3:00 sharp, he was back in the truck, ready to pick her up. Known around our small town as "Mariah's little prince," he could be seen attending her beauty pageants, her parent/teacher conferences, even her piano lessons. And there was that cute little trot of his, front paws turned out. His trot reminded me of our neighbor's Paso Fino horses: refined and ever so proud.

Mariah and L.A. went from small-town couple to America's sweethearts when they appeared on the talk show hosted by Leeza Gibbons, during a segment devoted to the raging debate over whether Pit Bulls were suitable pets. Six years old at that point and wearing a purple-flowered dress, Mariah kept her cool under the hot lights, and so did the snow-white Pit Bull. In fact, they seemed lost in their own little world as they sat on the couch next to me, kissing and giggling. More than a few in the audience who had come as haters left with a new attitude, thanks to the best goodwill ambassadors on the planet.

By that point, I had become embroiled in issues involving the shelter system and animal-related laws in general and was being asked regularly to conduct training and seminars all over the state. In 1998, amidst rumblings of a mandatory breeding license for Los Angeles–area dogs, it seemed time for "Beauty and the Beast" to make another public appearance.

The occasion was an open meeting of the city council in West Los Angeles. More than three hundred people showed up, half of them angry dog breeders who didn't think they should have to pay for a breeding license. The other half were rescuers like me, saddled with a staggering number of unwanted dogs and puppies, many of them "leftovers" from the breeders. Each side had come prepared to present its case to the council, and emotions were running high. At one point, the police were called in to restore order.

I bided my time. I had my own secret weapon to unleash.

When, at long last, my name was called, I sat still as a little eight-year-old girl and a white Pit Bull marched to the front of the room. L.A. seemed to know he was in the spotlight and trotted proudly, his head held high. When Mariah reached the microphone and realized it towered over her by about two feet, she quietly pulled an empty chair over to the podium, climbed up on it, leaned into the mic, and said, "L.A. . . . sit!"

With adoring eyes, he complied. Supporters smiled while the opposing team shook their heads, fearing they were doomed.

Just as we'd planned it together, Mariah introduced

herself and mentioned who her mom was, then talked a little bit about L.A. and the circumstances of his rescue. Her three-minute speech ended with, "I don't have to see the miracle of birth because every day I see the tragedy of death."

The room was silent as three hundred jaws dropped to the floor. A few seconds later, the place exploded in applause as my old soul of a daughter and her shopworn pooch strutted back down the long middle aisle to their seats. Though it would take some time to play out, the little scar-faced Pit Bull and his pint-size partner in crime had made their mark. In 2008, the Los Angeles Spay & Neuter Bill passed.

As Mariah grew from a child to a teenager, L.A. attended many sleepovers, birthday parties, and backyard camping trips, my daughter's constant protector on a wide range of adventures. One late afternoon, that four-legged knight in shining armor would be forced to prove once again how courageous these little dogs could be.

We lived on a twenty-acre parcel of land backed up against a canyon, with hills and some pretty rough terrain. Because my kids had been raised as country bumpkins, they had a knack for finding their own forms of fun. One day at dusk, with a very heavy fog rolling in, I was outside doing my normal ranch work when I heard a distant cry for "Mommy!"

My heart dropped, then returned to my chest, racing a

mile a minute. There was no doubt it was Mariah's voice I'd heard, and she clearly wasn't playing around. Her cry seemed to emanate from high up on one of the surrounding mountains, but the fog was so thick I couldn't see more than ten feet in front of my nose. I called out to her and she answered, sounding panicky.

Just then, one of the neighboring kids ran over and told me that Mariah, summoning her twelve years of wisdom, had decided to take my Suzuki Samurai and go "four-wheelin'." The fact that she'd learned to drive at an early age—and could even command a stick shift—was not un-usual out where we lived, but this was *not* a good day for her to be testing her off-road skills.

I began to make my way up the mountain on foot. It was extremely steep, and I climbed for what felt like forever. The fog was suffocating, and a chilly dampness stuck to my skin as I tried like hell to press on, one foot in front of the other, yelling out to Mariah the whole time to "just stay put." A vision of that car popping out of gear and rolling over a precipice kept flashing through my hysterical mind, but with each heavy step, I would shake it off.

When I could finally make out the little white Matchbox car through the soupy fog, I saw my daughter's sweet face peering anxiously out the driver's side window. There, next to her, was L.A.

"See, Mommy?" she said bravely when I came into view. "L.A. was here. I'm okay!"

As mad as I wanted to get, just hearing her sweet pixie voice—and knowing that the little dog had stayed by her side every second—made me break down in tears. If only human boys could be that sweet and loyal, I wouldn't have all those *other* worries ahead of me. The two of them climbed out of the car and we made our way back down the mountain on foot, still surrounded by a blanket of cold mist. Mariah babbled her apologies and promised to never try anything like that again.

Until next time, I thought, clutching her hand tightly in mine.

Seemingly overnight, my preteen turned into a full-blown teenager. A computer soon replaced the piano lessons and—inevitably—boys started to come into the picture. Mariah began to go on dates and to parties every weekend while L.A., her first and most steadfast boyfriend, waited faithfully for her to come home. Our house sat about five miles up a long and desolate dirt road, but he could sense her approach long before the headlights of her truck appeared in the front window. They'd been together for twelve years now, and L.A. understood my sixteen-year-old better than I did, continuing to love her fiercely and unconditionally in spite of having to share her with human boys.

The young lady and the Pit Bull still "spooned" at night and woke up together each morning to eat cereal and watch cartoons. Bath time changed to shower time, and L.A. patiently sat under the stream of water waiting for Mariah to finish washing her hair. He seemed resigned to her comings and goings, and the quick peck on the forehead she gave him as she walked out the front door. One of these two was growing up, the other growing old, but they remained as close and in tune as two creatures could be.

My little princess had met her prince when they were both four years old, and now, here they were, "adults." Just watching them curled together in sleep made my hard old heart melt. Pit Bull rescue had become my life, and although it was my passion now, it was also the hardest thing I had ever done. Most of my days included tears of frustration and anger—yet watching L.A. and Mariah together gave me the strength to keeping fighting the fight.

"Sweet 16" is every young girl's dream day. It's supposed to be filled with music, friends, and maybe even some pretty great gifts. But our family is cursed. Bad things just seem to happen on good days, and this day was no different. We had been noticing some nickel-size bumps on L.A. After taking him to the vet for a biopsy, it was determined that L.A. had cancer. For the next two years, L.A. would make several trips back and forth to the vet to have new tumors removed. With each trip, a piece of Mariah's heart would go with him. Putting him under to perform surgery at his age was getting risky.

We kept holding on, acting quickly when new lumps appeared and hoping no new lumps would appear, but acting again when they did. The night after each surgery, Mariah would gaze unblinking at the sky outside her window until she saw a shooting star, then send up a heartfelt wish for the health of her beloved friend. At eighteen, she'd started to stay home more, having discovered that the "boy next door" wasn't quite as cute or loving as the one who lived in her bedroom. She spent many evenings curled up on the couch in front of the fireplace with her little prince, speaking softly to him about the times they'd shared. There was the time she snuck him into her classroom and hid him under her desk. That memory made them both laugh out loud! And how about the time that L.A. showed up in a limo for their appearance on *Leeza*, decked out in a doggie tuxedo and looking every inch the sophisticated gentleman. The memories kept flowing, along with a river of bittersweet tears.

In August of 2009, L.A. died in Mariah's arms. He'd kept the cancer at bay, but the eighteen-year-old heart that belonged only to her just couldn't hold out any longer. As they sat in her truck together, she cradled him and kissed him one last time on the forehead. The angel of the city, her knight in shining armor, her prince, had bid a final farewell to his princess.

3

DUKE

The Action Hero

Unlike most young girls, I didn't grow up boy crazy—more like horse crazy. So once my fantasy affairs with Donny Osmond and David Cassidy wore thin, there was no doubt that I would marry one of my horses: either Benny or Cody. The Arabian geldings adored me, and each day after I'd finished my creative writing homework (another passion), we'd spend countless hours roaming the hills of Chatsworth, California, as I pretended to be an Indian squaw on a quest to free my people. Mind you, I have not one drop of Native American blood in me—but it felt intensely romantic to ride bareback and barefoot over the rocky hills of the West San Fernando Valley, my braided hair whipping in the wind.

I had a very active imagination in those days. My bedroom walls were plastered with images of wolves and African Painted Dogs. Eventually, the magnificent Triple Crown winner Secretariat found his place there too. These were my heartthrobs.

Thinking back about my childhood, it isn't surprising that I trusted animals more than humans and preferred their company. My biological mother took off when I was just a baby, and I watched my stepmother get emotionally trampled by my father for years. She had always suspected him of cheating on her, but I'm the one who walked in on him and saw the stark truth with my own eyes. At fifteen, I broke the news to her that she wasn't crazy: Dad was every bit the scoundrel she thought he was. All hell broke loose in the house, and there was no escaping the sound of my parents going at each other day and night. I found myself withdrawing even more fully into the world of animals as a deep distrust of people—especially men—took root in my heart.

This time Dad's the one who left, and for the next two years, it was just my stepmother and me. Although we didn't share blood, she was "Mom"—a tough lady, raised as a country girl herself—who taught me not to depend on anyone . . . and certainly not a man. She worked two jobs to support us and our animals, and at night I'd hear her crying in her room until she fell asleep to the Gatlin Brothers' "She's a Broken Lady."

Eventually her pain turned to anger, and she began directing it at me, the only link she had to the man who'd ripped her heart out. By the time I turned seventeen, our relationship had become toxic. I packed up my two horses, my Catahoula/Pit mix named Cougar, and my pet goat, Pete—my saviors and confidants—and we hit the road.

Without any semblance of a plan, I managed to drift for several years from one friend's couch to another, all the while keeping my animal family together. Despite my new-found freedom, I couldn't help but ache for Mom, knowing the pain she must be in back home, all alone.

I started to create pretend families in my head, always casting myself as the self-sufficient matriarch. I knew I wanted to have a daughter someday. I would name her after my idol, country singer Tanya Tucker, whose bold and outspoken lyrics gave me strength when I needed some. I vowed that I would do everything in my power to ensure that my own kids would never have to go through what I did as a teenager; I would be the only parent they'd ever need. In the words of a Tanya song, "I'll Come Back as Another Woman."

Years passed. Four kids later (one of which became Tania), Villalobos Rescue Center was off to a good start. I had earned the respect of the animal community with my background in wolves and wolf/dog crosses—which, at times, could be more difficult than their purebred cousins. Animal shelters across the state called on me to take Pit Bulls, and at the time I felt honored to be entrusted with this still controversial breed. I was dealing with two of the world's most misrepresented canines.

On June 12, 1995, when I got the call from the Pasadena Humane Society to pick up an "interesting" dog, I was happy to help them out. At the time, Pasadena did not adopt out Pit Bulls directly, but they sometimes called on me to assess them and potentially find homes for them through Villalobos.

As I made the one-hour drive to Pasadena, I mulled over what they'd told me about the animal, and it made me a little uncomfortable. Apparently, he'd shown no sign of emotion. Then as I walked through the shelter with one of the employees, I could tell that he was trying to sell me on this dog. "He may look scary," the guy said emphatically, "but he's very well behaved, and I'm sure that once he gets out of here, he'll be better."

I was preparing myself for some sort of saber-toothed tiger, but I was also intrigued. As we rounded the corner of the kennel area, there he was.

His name was Duke, and he was lying impassively at the back of his kennel as we walked up. He was a large dog, red—a color for which I had a particular fondness—with professionally cropped ears, as opposed to the street thug chop jobs I was used to seeing. He had clearly been well cared for and didn't sport a trace of that "junkyard dog" look.

The kennel worker explained that he had been dropped off by his owner because he wasn't a good enough guard dog. I'd heard that excuse before and couldn't help letting out a sardonic laugh. To think that this or any dog had been dropped off for not being mean enough pressed all of my buttons. It also changed my expectations.

I called Duke to the front of the cage, and he slowly got up and just stood there unblinking, as if analyzing me. And despite the fact that his size and cropped ears lent him an intimidating look, I didn't get a bad vibe from him. We locked eyes for a few minutes, mutually hypnotized and ignoring the sounds of barking dogs and the drone of the kennel worker still making his best used car salesman pitch. Finally, Duke sauntered closer and gave me a slow, nonchalant wag of his tail. Then he stretched, slowly, lay back down, and looked up at me as if to say, "Well, now what?"

My heart began to go pitter-patter for this bad boy.

By this time my soul had been torn to pieces by men twice over, and despite the fact that each of these relationships had gone on for a number of years—and resulted in my precious offspring—there were serious problems with both. I know, I know . . . I'd promised myself as a teen that I would never trust anyone too much, but I'm only human. I had succumbed to love. I'd given it the ol' college try. And after each bitter disappointment, the same Tanya Tucker song popped up on the radio in my head: "It's a little too late to do the right thing now."

After all that, I was finally *really* ready to look out for myself, my kids, and my dogs. I had no intention of adding anyone to my pack, human or otherwise. Then along came this canine, this man of a dog that almost instantly made me feel safe and secure. I just couldn't help being attracted to him.

★ ★ ★

Duke was a natural athlete. He could run like a gazelle and soar over a fence like a mountain lion. Although I was heavily involved in training at the time (training dogs for the movie industry), Duke's ability to learn new things was astounding, even to me. It was his thirst for learning and his eagerness to show off his talents—as well as his desire to please me—that helped me land what would become my favorite gig of all time.

Duke and I spent every second of every day together, and he always seemed to know what my next move would be. It was as if he could read my heart as well as my mind. It may sound sentimental, but although we were of different species, we were completely in tune. We slept together, traveled together, and loved to work side by side. So when I was approached by a friend who ran a company that trained animals for the movies, I jumped at the opportunity to get involved in a movie that called for a Pit Bull as one of the stars.

The movie was *Break Up,* and a Pit Bull was needed to share scenes with a woman (played by Bridget Fonda) who was running from her abusive husband. The dog would have to learn many different "behaviors," estimated to require about three weeks of intensive training. This would be followed by two months on location. Although I had already worked on one movie overseas at that point, this one would be much more stressful in some ways. For one thing, I would not have other trainers to rely on for help. It was just Duke and me this time, and we had to have each other's back. Yet . . . I could almost immediately envision it as a new career, and I dug in hard.

Would Duke have the right stuff to perform on cue? His past was a complete mystery, so I had only blind faith and our deep connection on which to rely—but I believed in him all the way. He'd proven himself to be sharply attentive, even intuitive. My big "career move" would be a golden opportunity for my failed guard dog as well.

Because some of the stunts Duke would have to perform were quite difficult, his success would depend on the bond he formed with Bridget Fonda, as well as our own working relationship. The three weeks of "prep training" were grueling for me, but Duke never seemed to get tired or lose his cool. Always a consummate professional, he remained in complete control, learning each new behavior after just a few tries. Watching him in action, I felt a little bit like a high school cheerleader watching her quarterback boyfriend make a touchdown. There wasn't a stunt he wouldn't try, and he refused to stop practicing until he'd mastered each one to perfection.

One of his most difficult stunts involved running across a busy street while dodging traffic, then jumping up onto the hood of a car and barking at the "bad guy" through the windshield. This took a combination of trust in the stunt drivers—some of the best in the business—and reliance on my training. It was up to me to make sure Duke would be safe but also get the stunt done. The truth is, this dog was so confident that he needed no encouragement beforehand or validation afterward. With the camera crew in place and me in the "bad guy" car, Duke was ready to show everyone what he was made of.

As the stunt driver maneuvered the car through the intersection, Duke ran into the carefully orchestrated traffic.

I watched nervously through the windshield as cars screeched around him, but Duke hit every one of his marks, then charged toward us. I couldn't help but get swept up in the moment as this Adonis of dogs pounded the pavement, muscles rippling with each stride. I gasped as he leapt up onto the hood and we came to an abrupt stop. His performance was literally breathtaking.

With a little direction from me, Duke then raged at the stuntman standing in for the villain, his saliva spattering all over the glass. With each command to "Speak!" he lunged at the windshield so viciously he made the camera guy jump. Duke was an artist, a true performer, and he clearly loved the spotlight. By the end of the three-week shoot, he had secured his place as the Lassie of Pit Bulls, and we were on our way to a joint career in film and TV.

There were more movies, as well as commercials and music videos. Each gig offered Duke a new challenge, and he was always up for it. It seemed there was no stunt we couldn't pull off—even when a director or a producer decided to change the action midstream, putting us to some serious tests. For instance, there was the time we worked on a movie out in the blazing hot Nevada desert alongside a couple of young and rising stars named Jared Leto and Jake Gyllenhaal, in a movie called *Highway*.

The scene called for Duke to sit in a chair as Jake's character and another man talked about drugs, alcohol, and women. In a moment that was supposed to be funny, Duke was to suddenly jump up and grab hold of Jake's arm and pull his jacket off. I knew this would present no problem

for my skilled stunt dog. We prepped it and nailed it on the
first take. After the director yelled "Cut!" he approached us
and stood for a full minute, his hand cupping his chin, deep
in thought.

"Okay . . . I have another idea," he said. "I want the
same basic scenario, but this time I want the dog to take
Jake's jacket off by the sleeve and drag it down the hallway,
around the corner, and into the bathroom." He said this as
if he were asking a stagehand for a glass of water. Like it was
the simplest request in the world.

I held it together for a moment, but once outside, the
suffocating Las Vegas heat made me angrier and angrier.
I began to cuss out the director. Well, not to his face. And
as much as I didn't want to admit it, a bit of panic was
beginning to set in. This was not what we had signed up
for. We had prepped for the action called for in the script,
and to think that Duke was like some sort of robot that
could be programmed . . . I was pissed. As I threw my hissy
fit for about five minutes, I looked over at my van. There
sat Duke, watching me. He was calm, showing no emotion
except within those Hershey-chocolate brown eyes of his.
They had such a calming effect on me; their warmth melted
me on the spot. As a slight smile began to emerge from my
now-chapped desert-dry lips, Duke's tail did that slow tail
wag that always said, "Baby . . . I got this." And then I knew:
We got this.

I confidently marched back into the house where we were
filming. Although I was just throwing this together, I knew
it would work—and I needed the director and entire pro-

duction company to believe it too. My plan began to come together as I laid out the tools necessary to do what would become one of Duke's greatest film-training accomplishments.

Remember, "all" Duke had to do was yank a guy's jacket off, then run down a hallway, turn a corner, and run into a bathroom. The problem was that I couldn't be right there to direct him or I'd end up in the shot. And I needed to be able to communicate with him throughout the process, step by step.

A bold idea was called for, something genuinely creative (if not insane), and that's what I came up with. I laid some walkie-talkies supplied by the production company all along the hallway, one hidden behind a plant, another under a rug, et cetera. Another trainer would stay with Duke to hold him in place until "Action!" then the dog would have to run the gauntlet as I directed him remotely. Duke wasn't accustomed to following my voice without seeing me, but I thought he could pull it off. I would be waiting for him in the bathroom, where I could call him into my waiting arms. I crossed my fingers and toes, nodded confidently to the director, and silently hoped for the best—that the gambit would work.

When everyone was in place, I told Duke to "Stay!" and ran down the hallway to the bathroom. My training helper then held the first walkie-talkie up to Duke so he could hear me yell out, "Duke . . . *get your toy!*" (This was the command we used when we wanted him to "Attack!"; we'd hidden his toy in the sleeve of the stuntman standing in for the actor.)

Duke followed my command instantly, as if I'd been standing right beside him. As canine and human struggled (for the cameras), Duke poured it on, tugging and thrashing his head about as he clung to the jacket sleeve. Since I couldn't see from my hiding place, I relied on the other trainer to fill me in on the action via walkie-talkie.

When the battle had raged on long enough to satisfy the director, he cued me and I yelled, "Bring it here, Duke! C'mon, buddy, bring it to me!"

"Okay," the assistant told me, "he's got the jacket and he just jumped down off of the chair . . . he's bringing it down the hall . . ."

Damn, this is some stressful work—but maybe we'll pull it off, I thought. Then . . .

"Oh, crap! He just stepped on the jacket and dropped it . . ."

I could hear the tension in the assistant trainer's voice, and of course my knee-jerk reaction was to jump up and help Duke—but I couldn't. The strength of our bond was about to be put to the test.

"Duke, buddy . . . pick it up. C'mon, pick it up!" I said firmly into the walkie-talkie. I waited for a few long seconds, and just as I was about to give up and emerge from my hiding place, I heard an excited voice.

"He's got it! Oh, my god . . . he picked it up!" said the trainer triumphantly.

I jumped to my feet and began to yell like a parent at a little league game: "C'mon, buddy, bring it here, you can do it! Follow my voice, buddy, just bring it here!"

Within seconds, my beautiful amber pooch was barreling down the hallway and into the bathroom, dragging the actor's jacket exactly as "scripted" (which of course it wasn't). He leapt into my arms to the sound of the crew all over the set exploding into applause. The director yelled, "Cut!" and then, "Great job, Duke!" I was so proud of my pound hound at that moment that I couldn't bear to let go of my chokehold on him. He truly was the Lassie of Pit Bulls, and a master of his craft like no other I've known before or since. Truth be told, I was in awe of him.

Duke became a constant in the film industry. Although he got bumped from Jennifer Lopez's music video for "I'm Glad" by my other lug of a dog, Moose—more about that later—he can be seen in any number of other videos, TV shows, and movies. Duke inspired respect wherever we went for his professionalism, quick thinking, and confidence, and the two of us expanded our mission by holding training workshops, seminars on dog safety, and other educational presentations. He continued to ace every new behavior I taught him within a few tries, sometimes even outthinking me and figuring out a stunt before I did. They say that "couples" shouldn't work together and that the stress of a job will bust up the best of relationships; Duke and I proved the exception to that rule. Our bond grew stronger with each new job we conquered.

★ ★ ★

Now, I don't want you to think that Duke was my only "significant other" throughout that time: I had been in a human relationship as well, though it wasn't what you'd call conventional. Mariah's father, Jon, with whom I'd been involved for twelve years, was a nice guy—but, sadly, he and I had different ideas of what a relationship was supposed to be, and somewhere along the line, those differences had become irreconcilable. At the breaking point, I moved out of the house and lived in a single-wide trailer on our ten-acre ranch, along with Duke and my other dog, Joe. Jon stayed in the main house and the kids went back and forth between us.

On those nights when I found myself crying on the couch, it was Duke who gave me comfort and the strength to hold things together for the sake of my kids and the dogs we continued to rescue and fight for. Whether he was resting his beautiful head on my legs as I slept or jumping in front of me to keep me out of harm's way, Duke was my constant protector, always appearing when I needed him—sometimes seemingly out of thin air.

Jon and I shared the property for a while, bonded by our love for animals as well as the kids. But there was one creature on whom we just couldn't agree: the dreaded Rex, Jon's pet rooster. Rex adored Jon but hated—and I mean *hated*—everyone else. This beast of a bird would come at all of us, claws spread like daggers, ready and willing to rip us to pieces. When we pleaded with Jon to lock up the big Rhode Island Red, he just laughed and called out to his "little buddy," who would jump obediently into his lap for a stroke on the head.

One busy Saturday afternoon, I had a couple of volunteers up working with the dogs and Mariah had a group of her six- and seven-year-old friends over, running around the property, screaming and yelling as kids do. Jon was in the barn with his horses. I opened the sliding glass door on my trailer to call out to Mariah, but she couldn't hear me over the soprano shrieks of her friends. As I stepped off my makeshift plywood trailer porch to move closer to her, the bottom step creaked and—just as if it were a scene in a low-budget horror flick—I froze. I knew "he" was out there somewhere . . . waiting.

I grabbed my trusty leaf rake just as I heard a wild squawk and saw a flash of feathers above me. Rex torpedoed down from the roof and landed squarely on my head, knocking me to the ground and tangling his razor claws in my usual mess of a hairstyle. I flailed wildly in an attempt to grab him and keep his beak away from my face, which was now covered with hair and dirt. The tussle seemed to go on for several minutes and then—with another flash of red—it was over.

I picked myself up from the ground just in time to see Duke running off and away from me. Then he stopped and turned back. It was then that I noticed ol' Rex hanging from my savior's jaws. Duke gave me a look of certainty and absolution; in that split second, he told me I was safe now and he would handle everything. And although I don't condone the killing of animals in this context, I understood that Duke was acting like any man protecting his woman. He'd come to my defense and handled business. And yes, I was somewhat saddened by the death of the "little bird," but I couldn't help feeling a sense of relief as my badass man trotted off into the bushes and dumped the body.

Together, Duke and I convinced Jon that his feathered sidekick must've gotten "eaten by coyotes." What happens in the desert, stays in the desert.

In 2004, Duke and I were on a "work high," having landed two major movies based on true-life events. And since my big red dog was in constant demand, we had built up a team of alternate look-alikes to serve as his stand-ins or even stunt doubles. Lefty, Moose, Gotti, Kato, Zeus, and Brownie each had a specific talent to bring to the table. Lefty and Moose, for example, were my "contact" dogs, so calm and chill that they could hold a "stay" practically indefinitely. I used them to fill in when Duke got a little antsy or looked too "edgy" through the camera lens. He continued to handle most of the action work, along with Zeus and Kato.

When I got the call to work with the well-known director (and dog lover) Nick Cassavetes, I was beyond excited.

His film *Alpha Dog* was based on events that had happened in the Los Angeles area. In a drug deal gone very bad, an innocent teenage boy was kidnapped and murdered.

The cast included such luminaries as Justin Timberlake, Sharon Stone, and Bruce Willis, and I knew this was a huge thing for us. But even more interesting than the opportunity to work with big-name stars was the fact that the father of the real-life drug dealer was hired on as a "technical consultant."

At the time shooting began, the drug dealer and suspected killer, whose name was Jesse James Hollywood, was still on the run. He was on every Most Wanted list and every bounty hunter was . . . hunting him. Seeing his father saunter around the set was unsettling, to say the least. We all suspected he must know more than he was saying, yet there he was, strutting around like a celebrity. I couldn't have known at that point that my run-in with Jack Hollywood would lead to a movie-worthy story of my own.

About two weeks into working on *Alpha Dog*, I got a call to work on another movie where some red Pit Bull action was required. It was going to be a juggling act, but I decided we could handle it, and on a day off from the first picture I headed out to the Mojave Desert to the location of the second.

What are the chances? It was another tell-all, this time the story of Domino Harvey, a model turned bounty hunter being played by Keira Knightley. Mickey Rourke was her costar. The set was an abandoned house in the desert and the shoot was at night, so, as anyone who understands that terrain knows, it was freezing cold. I settled my dogs in the van and joined the rest of the crew, standing around a burn barrel trying to keep warm.

Making a film involves a lot of waiting around. Some people read or play cards while others, like these people, just hang out and shoot the sh*t. Fine with me; I was happy to make some new friends. Intrigued by the deep laugh of Mickey Rourke, I introduced myself as the dog trainer and he immediately launched into his love of dogs. After we chatted a bit more, he jogged back to his trailer and brought out his pack of Chihuahuas to meet me.

Not much leaves Tia Torres speechless, but there I was, mouth hanging open, trying not to look surprised. One of the three, the littlest, was wearing a pink sweater with a skull and crossbones on it! My first thought was, *What is this big scary-looking dude doing with all of these teeny-tiny dogs?* As caught off guard as I was, I did think it was kinda cute . . . in a sexy kind of way.

After I met and made a fuss over each of them, Mickey and I got back to our dog talk, and I mentioned that I was also working on *Alpha Dog*. At that, another guy who had obviously been listening in turned to us and said, "Did you say *Alpha Dog*?" His muscled arms were a mass of tattoos and he was wearing a black leather jacket, but the tone in his voice could only be described as "excited panic."

A bit taken aback, I responded, "Um . . . yes. Why?"

In one continuous move, he put his finger up for me to "hold on a sec" and grabbed his cell phone, quickly walking away while punching in a number. I looked back at Mickey, who shrugged and picked up our conversation without saying a word about it. A few minutes later, the man came back and asked if he could talk to me in private. Reluctant but intrigued, I stepped off to the side and—regretfully—left Mickey Rourke to keep himself warm.

The man with the stature of a Bulldog introduced himself as Zeke, and said he was a bounty hunter from the Los Angeles area. He was very excited and talked fast. He questioned me about working on *Alpha Dog* and asked whether I was close to anyone on the set. I explained that I was the dog trainer and that the only cast member I had spoken to was Justin Timberlake—about his Boxer's skin condition. Zeke then dropped a bombshell on me.

"What about Jack Hollywood?" he persisted. "Have you had any contact with him?" I guess I wasn't surprised that this was where the conversation was headed. Of all the people I'd met on the shoot, he was the one for whom I had the least respect—and I told Zeke that. I explained that I'd had no personal contact with the man and was happy to keep it that way.

The story that Zeke then unloaded on me made my head spin. His bail bond company was one of many trying to find Jesse James Hollywood, and everyone knew that his father was very close to him. The rumor was that Jack had some secret way of communicating with his son, but so far no one had been able to crack it. Zeke and his partners wanted to know if I would be willing to get close enough to Jack Hollywood that I could help them find Jesse.

My mouth dropped open. I knew what it felt like to work

on movies but now I felt as if I was *in* one. Zeke's plan was that I figure out a way to get Jack's cell phone and use it to dial some secret number, making it possible for them to track any subsequent calls he made. I admit, it sounded exciting, but this was a lot to take in on the same night that I'd met Mickey Rourke. As dangerously tempted as I was, I opted to simply remain a dog trainer and let the wheels of justice roll on without me. And, as it turned out, my "assistant bounty hunter" duties were not even required: Jesse James Hollywood was eventually caught without my help.

Still digesting the events of the previous hour, I walked back to my van and found Duke ready for his big moment, as he always was. His task was to protect the character played by Martha Bozeman from being shot by Mickey and his bounty hunters, who were attempting to bust into her mobile home. Everyone was in place and I had another trainer positioned near Duke and Martha so I could hide across the room and cue him when the guys burst through the door.

Right before a director calls "Action!" the anticipation is always intense. The experience is especially nerve wracking when the scene is a complicated one, like this one was. It involved fake gunshots and chaos, and I knew I had to stay focused on Duke every second. Everyone was in place and the air was practically vibrating . . . when my fellow trainer called out, *"Wait!"*

There was a collective exhalation of breath and you could almost hear the silent moans of "Are you kidding me?" The director glared at me and I felt my face getting hot. Apologizing as I sprung up, I darted behind the kitchen counter to where my other trainer was squatting, about three feet from Duke and Martha.

"Duke just lunged at me," he whispered.

"What do you mean, 'lunged'?" I said, uncomprehending. "Aggressively?!" I was trying to keep my voice low enough that no one else could hear us.

"Yeah! All of a sudden, he just came after me! If Martha hadn't been holding him by the collar, he would've nailed me."

I looked over at Duke, who sat patiently, waiting to do his job. Martha stroked his head and, as usual, Duke showed no emotion, which was what made him such a great working dog. He was there to do a job, and he didn't get involved in the drama surrounding him. In all the time we'd been working together, he'd never acted aggressively toward anyone on set. For that reason, I just couldn't help doubting my training assistant's story. But, not wanting to take any chances—or waste any more precious time—I decided to do the stunt with Duke by myself. I knew he was good enough at his job that he really didn't need another trainer to tell him to "Stay!" After releasing the guy with a quick apology, I went back to my place across the kitchen and once again we waited for "Action!"

Mickey and the others came barreling in, guns drawn, yelling and making the requisite ruckus. Martha, a great dog person in her own right, held on to Duke's collar as instructed as I commanded him from across the room to "Speak!" He performed on point, barking and frothing at the mouth so much that spit hit me from across the room.

And when he had completed the take, Duke was rewarded with his favorite toy, a worn-out black spongy baseball shin guard that he loved to throw up in the air and catch. It never failed to astonish onlookers when Duke went from vicious killer to goofy puppy dog in a few split seconds. Actors were always particularly impressed. I guess they understood that he was one of them, and a great one at that.

Over the months that followed, little incidents occurred. Every now and then, Duke would snap at someone or move toward him as if he might bite. Of course, I kept making excuses for him, but I couldn't help but think about what had happened on the set of *Domino*. It got to the point where I started using one of my other red dogs in his place, insisting that Duke was just having a "bad day."

Inevitably, Duke himself started to notice the change. He became more and more depressed and just didn't exhibit his usual regal confidence. He was getting up there in years—he was almost twelve, to be exact—and though he still looked great, I could tell something wasn't quite right. With great concern, I arranged an appointment with the vet. And for the first time in all of our years together, my rock, my life source, the man of my dreams and I had to be separated for a night while they conducted tests on him.

It was the longest twenty-four hours of my life. But as I sat in the waiting room, anticipating the moment when Duke would come strutting out to greet me, I just knew everything would be fine. Dr. Mark Hohne would have solved the mystery and prescribed some medication and we'd be back in business. But instead of my partner walking through the doors, it was the vet—with a fistful of X-rays. His demeanor was serious, and his voice cracked as he began to talk.

"Tia . . . there's no easy way to say this. It's cancer."

Knowing how I felt about Duke, and having been our vet for years, even the doctor was fighting back tears. He began to explain the situation in detail but I couldn't hear a thing. The room was spinning and a wave of nausea washed over me.

I choked it back down and said, "How bad?" Not that I really wanted to know.

Dr. Hohne gently explained it all over again: that Duke had cancer in his spine and a tumor near his brain, which explained his behavioral changes. At Duke's age, he told me, chemo just wasn't an option.

"Tia, as long as he's eating and still functioning, he can live his life normally," the doctor told me. "But I must tell you . . . it's only a matter of time."

Right then, the vet tech brought Duke out. In my fantasy of the moment, I'd hoped he would race over to me, knock me to the ground, and tell me in his own way, "They got it all wrong, baby!" Instead he sauntered over slowly, head hanging low. As I bent down to greet him, he nuzzled my face as if to say, "It ain't good, baby, it just ain't good."

I managed to make it to the car before I broke down, and then I sobbed all the way home.

Duke was a strong specimen of a breed known for its

resilience, so it took a while for the disease to get to him. Little by little, he began to show signs that his body was failing him—but he wasn't going to go until he was good and ready. He lost his appetite for his normal food, so I began to give him anything and everything he wanted in an effort to keep his strength up. I knew it was selfish but I didn't want to let him go, no matter how bad it got. When I noticed his head tilting to the side and his back legs getting weaker, I clung to my own denial. And with each sunrise, I would wake up thinking it was all a bad dream and that he would be lying next to me, waiting for me to open my eyes so he could slap me in the face with his big warm tongue. Within seconds, the reality would knock me senseless and I'd cry.

In June of 2006, on Duke's birthday, I made the horribly painful decision to drive him to the vet. He could barely walk and only intermittently recognized me at this point; the miracle I'd prayed for hadn't materialized. I knew it was time—that it was unfair to make him go on one day longer.

We made our last one-hour drive together in silence. When I helped him stagger into the last room he'd ever see, I was determined to hold it together. I wasn't going to cry and make it worse for my man.

I held him and whispered in his ear as Dr. Hohne injected him. As sick as he was, Duke still fought for life, and that's when I couldn't hold on anymore. I unleashed an otherworldly, heartbroken wail as I clung to his warm, limp body. It was over. His pain was gone, but mine was just beginning.

The vet tech tried gently to take Duke from me and I turned into a wild animal, kicking at him, screaming obscenities and holding on. I yelled at everyone to "Leave us the f*ck alone!" and I cradled the finest animal that had ever walked the Earth, rocking back and forth until I too felt lifeless.

★ ★ ★

It's been years since Duke left me. I tried to replace him with a human counterpart, only to have my heart ripped from my chest yet again. There have been a few dogs who have tried to take over the spot reserved for Duke, but through no fault of their own, the hole just couldn't be filled.

I'd come from a place of distrust. I learned early to believe only in myself and to never rely on anyone. Duke changed that in me. I leaned on him and knew that he would always keep me safe. He put my feelings before his own. He knew what I was thinking before I did. He possessed the one attribute I'd never expected to find in anyone, man or animal: absolute loyalty. Duke was one of a kind, and there will never be another that will even come close to his greatness.

To this day, my heart remains guarded. After Duke departed this Earth, I went through some pretty dark times. Sometimes I didn't even think I'd pull through. I'd lost my rock and had no one to keep me strong. But, looking back at it all, I see that everything happened for a reason. It was all as it was supposed to be.

Duke had been preparing me for a much bigger journey, during which I'd be forced to stand up on my own two feet and take charge again. I had allowed others to make me weak, but it was time to fight and take back what was mine. Like the goddess Artemis, who ran with her pack of dogs, I was to be that warrior again and honor what Duke had taught me. To fight for what is right and bite back twice as hard. I will remain strong and shed no more tears.

Thank you for saving me, Dukey Boy. You will always be the greatest man I've ever met, but . . .

I got this now.

4

MOOSE

The Reluctant Rock Star

I grew up on a ranch surrounded by horses and cattle and competed in every category of horseback riding that existed, from rodeo events to hunt/jump competitions. Big animals were my life growing up, so I figure that's why the wild beasts I found myself working with years later didn't faze me.

As a trainer in the movie industry, I found myself hanging out with tigers, wolves, and bears, doing everything from trying to catch an escaped and mischievous grizzly who thought rampaging through a village in Sri Lanka was "fun," to wrangling a pack of wolves who felt racing through the jungle was better than any dog park.

In the summer of 1996, while preparing for a movie job, I had no one to watch Tania after school. So, one day, I brought her with me to work; her babysitter was Nellie the elephant. Tania spent the afternoon riding Nellie around while I trained my wolves. As if she were in a scene from *The Jungle Book* (which happened to be the movie I was working on), my little desert Mowgli was living a life of which every child dreams. Actually, this was how I raised all of my kids.

Animals were the center of our lives, and everything we did was for them. No matter how small or big, we loved them all and gave our life for them.

From the outside looking in, I suppose the average person might have thought we were nuts—or that we lived an excitingly dangerous life. I mean, what mother allows her eleven-year-old child to run around on the back of an elephant? What sane person lets her children play in a wading pool full of monkeys? Maybe in some ways those people would be right—but this is the life we chose. While many kids were running the streets getting into trouble or joining gangs, mine were cleaning kennels and preparing the next day's meals for our menagerie.

In raising my kids as I did, I found it strange and interesting how social pressure truly worked. It was one thing to have my kids growing up around exotic animals. "Normal" people passed this off as: "Well, those animals are *trained professionals*, so it's okay." But if I showed up to drop my kids off at school with a Pit Bull hanging out the truck window, my next visit might be from child protective services. As an animal handler, I was embarrassed for these uneducated and foolish idiots.

So when I received a call from L.A. Animal Services in downtown Los Angeles to pick up a "beast," I was all too pleased to once again ruffle the feathers of the uppity and judgmental folks of my upper-class yet small-minded town.

As usual, I was greeted at the shelter with smiles and open arms. I had a great relationship with the local sheltering system, where for seven years I had been conducting free dog-training classes called "For Pit Bulls Only," open to anyone who owned a Pit or Pit mix. L.A. Animal Services had become very proactive in adopting out this misunderstood breed—so when they told me that this particular dog was one that I had to see for myself, I was more than a little bit intrigued.

As a shelter worker escorted me to a private area down many corridors and through several locked doors, I began to feel as though I were entering a scene from *Dexter.* I mean, what kind of beast could possibly have to be kept behind closed doors? But my escort's demeanor was lighthearted, so I was confused. "What's up with this dog?" I demanded, but he just smiled and said, "Oh, you'll see."

Why was this dog being kept so far from people and other animals, behind locked metal doors? What atrocities had he committed? My answer was waiting for me behind one more secret door.

There he was, the largest hunk of a lug I've ever encountered. He looked like a Pit Bull who'd been radiated in one of those science fiction movies. My escort informed me that he'd weighed in at a massive 110 pounds. No way he was purebred, but what in hell was he mixed with? He was well proportioned, with a perfectly chiseled bowling ball head and cheekbones carved from granite. His ears had been professionally cropped by a veterinarian and were perfectly shaped. His fur was the deep red color of Georgia clay. *I'd kill for hair like that,* I thought. His eyes were tinted gold—not that scary yellow color you see on Halloween masks—and his gaze was lazy and gentle, much like his entire demeanor. And his body—all I can say is, *Wow!* His muscles had muscles! But it was his body language that I found most appealing. Even in solitary confinement, with only the distant barking of his fellow inmates for company, he lay there, head resting easily on his front legs. His only acknowledgment of our presence was a slow wag of his tail.

"So . . . what's this guy's story?" I asked, impatient for an explanation at this point.

My escort explained that they'd locked the dog away for his own protection, not wanting him to "fall into the wrong hands due to his looks." I got his point: This guy—Moose, by name—looked every inch a "gangsta." But how did such a magnificent animal end up in a shelter in the first place?

Apparently, Moose's owner had surrendered him because he'd gotten "too big." Whatever. I will never understand people, and I really ought to give up trying.

"I'll take him," I said, as if there was any doubt.

Moose instantly fit right in with our gang at home. Just as he was at the shelter, he was "chill" as he integrated into Villalobos. He was this lumbersome, easygoing, goofy dog who just loved everyone and everything he met.

I was still working in the movie industry, and I thought Moose would make a great addition to my "red dog team." Although the biggest of them all, he would definitely have a place within my working group. He was not very energetic and couldn't do any stunts, but he had the looks to make up for it, as well as a solid temperament that meant he didn't seem to care what was going on around him.

With Duke as my lead movie dog, Moose had some pretty stiff competition. Duke could learn a command in a matter of minutes. Moose, on the other hand, would look at me as if I were speaking a foreign language. I tried everything: treats (both dog and human), toys, you name it, to motivate him to learn something as simple as a "Sit!" He would just stand there, stare at me, and drool. Then, after a few moments of being bored, he would lay down and go to sleep, leaving me standing there looking like a fool. Moose was a challenge. Nothing motivated him. We worked around his schedule . . . which meant allowing a lot of time for naps.

Whatever his shortcomings, Moose had a great role to play in my educational work. His calm friendliness made him ideal for school presentations, where he would lay sprawled out on the floor as kids lined up to hug him and play with his big floppy lips. Even his name seemed to delight them.

Moose was also a big hit with the members of our juvenile hall program. When that big guy strolled in, he immediately got the attention of the junior gang members, who assumed from looking at him that he was the ultimate tough guy. His gentle goofiness threw them for a loop, and it didn't take much to make the point that dogs like Moose were born to be lovers, not fighters. Whenever we needed a "gentle giant," Moose was our man. It never occurred to me that he might be a TV star as well, but, then, you never know what's around the next corner.

One day I got a call about a music video that specifically required a *big* red Pit Bull. The only thing the dog would be called upon to do was "hang out" with the singer and run alongside her bicycle. The producers sent over a photo of the kind of dog they had in mind, and indeed it resembled Moose to a T—but the action called for a dog with Duke's skills.

Curious about the particulars, I asked who my dog's co-star would be and the response had me giddy with excitement. Jennifer Lopez! Yup, J.Lo would be working with one of my dogs!

I tried to sound professional but was having a hard time of it while jumping up and down in place as the production assistant kept talking. The song was "I'm Glad," and the video would be a kind of tribute to *Flashdance*. That's when it hit me: No wonder the Pit Bull in the photo looked so familiar! The picture was a still from *Flashdance*, one of my favorite movies from the eighties! *I loved the eighties.* This was going to be a dream gig and the best thing to happen to Pit Bulls since Petey joined the *Little Rascals*.

I was so starstruck thinking about J.Lo that I almost missed the other bit of news the PA had for me: The director of the video was David LaChapelle, known to be one of the toughest in the business. That brought me back down to Earth quick. I realized I'd better get my you-know-what together if I was going to pull this off.

The first step would be to introduce J.Lo to her costar, so the very next day, I got Duke spit-shined and ready for his close-up, and I read the script they'd sent over to see what else might be in store. I knew Duke could easily pull off the task of running alongside a bicycle, but there were also a lot of long "Sit!"s and "Stay!"s while the action went on. To an outsider, that might sound easy enough—but the truth is, getting a dog to just *do nothing* can be the trickiest thing of all. That's especially true when a bunch of crew members are moving equipment around, actors are dancing and singing . . . you get the idea. Give me "Jump over that wall!" or "Hit that fence!" any day over "Stay!"

On any movie job, it's a good idea to have a backup dog available. In this case, I decided to take Moose along (despite the fact that he had *zero* training). He did look like the original *Flashdance* dog, after all, and I thought maybe J.Lo would like him. I'd explain that Duke would do most of the work (okay, all of it) but that maybe Moose could do a couple of close-ups for the camera.

I reached the filming location (which, by the way, was the original set of *Flashdance*) and tried to keep my heart from racing as J.Lo and the assistant director came out to my van. She was all smiles as she introduced herself and asked to see the "doggies." I told her about Duke's vast experience and how professional he was, boasting that they would probably get their shots in one take. Then I opened my van doors to let *the pro* out of his crate.

Imagine my surprise and embarrassment when Moose, the big dumb lug, tumbled out and landed on his head in the dirt. Apparently, in my excitement to meet a superstar, I hadn't closed his crate all the way. I quickly helped him up and apologized to a giggling J.Lo. She thought he was adorable and began to smush his face in her hands and make friends with him. I stammered out a quick apology and explained that Moose was just along for the ride and that the "real" dog was still waiting patiently in his crate. I'm pretty sure nobody heard me; J.Lo certainly didn't, busy as she was with Moose, who was melting in her arms like a lovestruck suitor.

"This is the one I want. He's adorable," she said, as she squished his face one more time before running back onto the set.

Who can say *no* to J.Lo? I was in trouble.

"It can't be him," I whimpered to the assistant director. "He really doesn't know anything—he's just a big dumb dog. Really—you have to let me use Duke."

The assistant director just shook his head and said, "Whatever Jennifer wants, Jennifer gets. And she wants that dog. See you tomorrow." With that, he walked away, leaving me standing in the parking lot with my mouth open.

I got back in my van, looked over at Duke, and said, "Buddy . . . we are so screwed." And, as in sync as a bonded couple could be, we both snapped our necks to look back at Moose. Without a care in the world, there he sat, his muzzle still covered in dirt from his face plant "audition" for J.Lo in the flesh.

I didn't sleep that night, and as I tossed and turned, all I could envision were leg warmers. Lots of leg warmers. It was going to be a long day.

★ ★ ★

Exciting as working on a music video sounds, it can be extremely stressful. Loud music goes with the territory, so the dogs have to be able to deal with that. My dogs were pros, of course . . . but Moose was an unknown quantity, so I was tense as I parked my van and started to set up. When one of my fellow trainers heard about the gig, he'd skipped right over the J.Lo news and said, "You're working under David LaChapelle?! Good luck with that one!"

Well . . . everyone loves dogs, I said to myself as I unloaded Moose, leaving Duke in the care of my senior trainer. I could see that Moose was turning heads as I walked him around to acclimate him to the location. He was easy to handle, as usual, and more than happy to greet every one of his admirers. I decided to take him over to the soundstage so he could get used to the music—which took us right past craft services.

Craft services is where the cast and crew eat, get snacks, and so forth. The budget of the job determines the selection— and this was a J.Lo music video. I'm talking, they had every kind of food imaginable. As I came up to the table, I noticed stacks of crab legs. Food would certainly comfort me for the time being, and there was nothing like relaxing to a belly full of such delicacies. At the very moment I reached for a plate, Moose decided he wanted in on the seafood platter too. With my plate full of legs in one hand and Moose's leash in the other, I tried to stop the train wreck from happening, but it was too late. All 110 pounds of doggie beefcake jumped up on the table. Moose's front legs caused the table to teeter on its side as frantic chefs, assistant chefs, production assistants, and others raced over from all corners to save the high-priced smorgasbord. Thankfully, the only casualty was my own plate, which splattered to the ground, causing more embarrassment than disappointment on the part of my tummy. As I apologized up and down and even sideways, I slithered away knowing that the news of the near disaster might reach the feared director faster than I could show my face on set.

The thought of switching out Moose for Duke came to mind. Would anyone notice? I mean, they were similar in appearance, so maybe I could slide him in and show off his talents to this David LaChapelle? These were my thoughts as I continued toward the production office, red face and all. I was yet to realize the extent to which I was in over my head.

Our first shot was the one where Moose had to run alongside J.Lo on her bicycle. We had prepped this back at home, and Moose had actually done okay. I explained to the assistant director that as long as J.Lo rode at a moderate speed, Moose could keep up. The location was a typical small-town street scene, and everything looked good until I noticed . . . the break-dancers! They wanted J.Lo and Moose to ride past this group of gyrating young men. This was unwelcome news to me, but I gave a "no problem" thumbs-up and a weak smile. Moose would no doubt want to stop and check out the dancers and—who knows?—maybe even join them. This was going to be challenging.

I positioned the other trainer down at one end of the

street with Moose and J.Lo and walked toward the camera, where the dreaded LaChapelle was about to yell *"Action!"*

From down the street, I saw J.Lo take off on her bike. At first I didn't see Moose.

"C'mon, Moosey! C'mon, baby!" yelled the star in her cute baby voice. *Hell,* I thought, *I'd run after that.* And . . . along came Moosey.

They sped down the street together, and when Moose fell a little bit behind, J.Lo slowed down just enough to keep him by her side. As they neared the break-dancers, I saw Moose start to drift in their direction, and my heart skipped a few beats. I had to think of something, quick. I stood up so that Moose could see me and began calling him as loudly as I could. I jumped up and down like an idiot, waving and yelling until I got his attention. As he began to run toward me, I was flooded with relief. Then I felt something against the back of my leg. Reluctant to take my eyes off of my rock-headed dog, I stole a quick glance behind me and saw Duke! Where the hell had he come from?!

In an instant, Moose was plowing me down, Duke was jumping on both of us, and I was still trying to figure out what was happening. Apparently, Duke had heard "Action!" from his spot in the van and thought it was his cue. He had busted out of his crate, jumped out the window, and assumed "Places!"—then danced around waiting for a reward.

To the sound of applause for Moose, I attempted to gather my frazzled self together and pretend I was in control of a very out-of-control situation. *Yeah, sure . . . this was how it was supposed to go. I meant for that to happen.*

As Duke continued to pose and preen, wearing one of those big 'gator smiles for which Pit Bulls are known, Moose had already tired of the attention and settled down for his nap. The scene, crazy as it had been, had left me a little more confident that I could pull the job off.

The next set was designed like an old New York loft apartment. In the center of the floor was a chair, and various props were strewn around to indicate that a dancer lived there. As the crew continued to arrange things, I was told that the director wanted to talk to me. *Gulp.*

As if summoned by the Godfather himself, I was escorted by a couple of crew members over to the guy in the big director's chair—where I just stood, tongue-tied, like a kid in the principal's office.

"You're the dog lady?" asked LaChapelle distractedly.

Still unable to form words, I just nodded.

"Okay, so here's what needs to happen. The dog just needs to sit there while Jennifer sings. She'll be sitting in the chair and he needs to do nothing. Absolutely nothing but sit there. Easy enough, right?"

Before I could answer, he got called into another conversation and walked away, leaving me standing there wondering how we'd pull off this "easy enough" trick.

As I mentioned, to "just sit there" is the hardest stunt in the book. Moose had never even pulled off a proper "Sit!" let alone a long "Stay!" There was only one thing I could do: I turned to Debbie, my senior trainer, and said, "We need to use Duke on this. There is no way that Moose can do it."

Debbie went off to try and convince the assistant director

to let us bring in the pro. Seeing that she came back way too quickly, I already knew the outcome. I could tell by the look on her face that she hadn't succeeded in her mission.

"He said that Moose did so good today that they want to keep using him," she told me. She'd done her best to explain that running alongside a bike is one thing, sitting still quite another—to no avail. Moose was in for better or worse.

I had to wake Moose up when it was time for the shot— which I hoped was a good thing. Duke popped up, ready to work, but I just kissed him on the nose and said, "Wish us luck, buddy. We're gonna need it."

Moose stretched, yawned, and stretched again. *If we're lucky, maybe he'll just lay there at the feet of one of the most beautiful and talented women in the world.*

Yeah . . . right.

All eyes were on Moose as I got him into position, and a few people reached out to pet him as we walked by. Others just stared blankly, waiting for the director's order. I stole a glance at LaChapelle's "throne" and found him hunched over, head in hand, as if to say "I'm bored, let's get this over with." Or maybe he was in a state of deep concentration; either way, the pressure was on, and I knew it. The heat from the stage lighting had become almost unbearable, and I felt faint as I stepped Moose up onto the wooden platform and positioned him next to the chair, quietly entreating him to "Sit!"

Jennifer came in and took her place next to him, and whether it was her perfume, her beautiful eyes, or that cute voice of hers, Moose was on the loose. He jumped up and became a puppy whom I didn't recognize. Charmed by his antics, she began to pet him and kiss his face.

This was becoming a dog trainer's nightmare.

When working with actors and animals, it's best if they don't have a lot of contact. The animal's job is to pay attention to the trainer, and playtime gets in the way of that. But this was J.Lo, after all, so I tried to intervene diplomatically.

"Oh, Moose," I cooed, "I know your new friend smells nice and looks beautiful, but you have to listen to your mama!" This was my gentle way of saying, *Please leave the dog the "f" alone!*

Finally, with the help of some bits of raw hot dog I kept in my treat bag, I managed to recapture Moose's attention. When he was sitting like a gentleman, perfectly focused, I stepped back and motioned for them to start the music.

The song "I'm Glad" exploded throughout the studio. It had a really cool beat, not too slow and not too fast, and Jennifer began making some *Flashdance* moves. As she got into the groove, Moose did what Moose does: He stood up and attempted to kiss his new friend.

"Cut!" bellowed LaChapelle from behind me, and even though I knew it was coming, it scared the crap out of me. Jennifer was giggling and playing with my silly dog, and most of the crew was laughing. Two people were not laughing: me and the very pissed-off director.

"All of you may think this is very fucking cute but *that* just cost us thousands of dollars!" LaChapelle bellowed, then looked directly at me. "Do we have this or not?"

Without saying a word (apparently, he was never going

to hear my voice), I tried to calm Moose down and get him back in place, but now he thought we were playing a game. I shot my trainer, Debbie, a look that said "Help me," hoping she'd come up with some miracle gimmick. Moose wriggled out of my grasp and spun around like a bucking bronco, and I could feel dozens of pairs of angry eyes on me. Sweat dripped down my face and burned my eyes. I was calling the dog every foul name in the book when Miss Lopez decided to "help"—by *playing tug-of-war using a ballet slipper!*

I began to panic. What was I supposed to do? Blame Jennifer Lopez for my dog's bad behavior? Instead, I did the unthinkable and asked the almighty David LaChapelle if I could (waste more of his time and money and) "take a moment" to regroup with Moose.

The assistant director came over and begged me to get anything I could out of the dog. "Even if he just hangs around . . . or sniffs the ground or something . . . we have to get this shot!" Of course, all Moose wanted to do was play with his new friend, the former Fly Girl. So Debbie and I hatched a very unconventional yet desperate plan.

We drilled a huge eyebolt into the wooden platform and clipped Moose to it. This was not just any old eyebolt but the kind used to anchor heavy equipment such as tractors and bulldozers. We crossed fingers, toes, and paws that it would work on a pigheaded, dumb-as-a-box-of-rocks, 110-pound lovesick beast affectionately renamed Moosey by his beautiful costar.

As slow on the uptake as Moose could be, it took him

only a few tries to realize that his bulldozer butt wasn't going anywhere, and he lay down next to Miss Lopez, who was still giggling with delight over his antics. Although I didn't share her amusement, I did think the scene might finally work and managed a small sigh as cast and crew assumed their places.

The director called "Action!" and the music started up, and J.Lo began singing. Moose just lay there, panting . . . until his new "boo" glanced over at him as she sang about love and longing. Then . . . *pop!*

The fact that Moose was anchored by the collar didn't mean the rest of him had to stay put. His butt came up while his head stayed down (as if he were bowing to the crowd) and he let out a *"Woo woo woo!"*

Jennifer burst into laughter.

I could see the veins protruding from the side of his head as LaChapelle yelled, *"Cut!"*

It would take another hour of my having a borderline mental breakdown, but after several more takes, one more ballet slipper, and one torn-up towel, we finally got the shot. Then LaChapelle had another "idea."

If you're a *Flashdance* fan, you remember the "Maniac" scene, where Jennifer Beals is exercising by running in place while her faithful (and much better behaved) companion watches. Well, our esteemed director decided he wanted to re-create that moment, and once again Moose would have to sit still for it. And—oh, yeah—because it was a workout scene, they wanted Moose to wear a towel around his neck! As in *the very same thing that J.Lo had just been using in their tug-of-war game.* I honestly wasn't sure if the man was trying to torture me or give me a chance to redeem myself.

Dogs are dogs. Their thought processes are not like ours. In Moose's mind, the towel was his toy and he could do whatever he wanted with it, including tear it to pieces. Every time I placed it around his neck, he reached around, pulled it off with his teeth, and shook it. We hadn't even gotten to the "Sit!" part yet. Trying to get Moose to leave the towel alone was like unringing a bell, and I could feel the director and everyone else getting restless.

I brought out the pieces of hot dog to get Moose's focus off the towel and onto me. Every time he managed to stay away from his "gym attire," I rewarded him—but the problem was that his attention span was that of a gnat on crack, so by the time I'd gotten him in line, I'd been shoving hot dogs into his mouth for what felt like an eternity.

Finally, he sat still, looking like a stud muffin with the towel draped over his broad shoulders. I slowly motioned to the director that we could start the cameras and, like the pro that he is, LaChapelle knew better than to yell "Action!" He just made a circular motion with his hand so as to not distract the "star."

Moose sat up on that pedestal like a champ. This was his moment to prove himself a genuine trained movie dog. Until . . . something began to happen. He let out a couple of "chuff"s while heroically holding on to his hard-won "Stay!" Then, like Krakatoa on a bad day, Moose's entire body began to heave. Making its way from the depths of his very large abdomen, up through his Hercules-size chest, and into his throat came the most horrific sound imaginable—like an alligator with really bad acid reflux—followed by an *explosion*.

I saw something catapult toward me and ducked just in time to avoid getting smacked in the eye. And before I could turn around, I heard King David scream at the top of his lungs, *"What the fuck was that?!"*

Slowly, I turned to face my execution, only to see a cameraman furiously wiping his lens with a rag. Apparently, Moose had ingested a few too many hot dogs and, like a sniper with dead aim, managed to upchuck with such force that a regurgitated bit of meat shrapnel hurtled through space and hit the lens dead center. Pit Bull's-eye.

We were excused and paid, and we left with our tails between our legs. Or I should say, *I* did. Moose, on the other hand, had had a blast and made lots of friends. As exciting as it would be to tell everyone that we made a J.Lo music video, I was relieved it was over.

But things have a way of coming back full circle and biting you right on the ass.

A few months passed, and one Sunday afternoon I was watching MTV—back when they actually played music videos. A show came on that I'd never heard of, called *Making the Video*, and guess what video they were featuring? Yep. "I'm Glad."

This should be . . . interesting, I thought, actually excited to see the cool stuff I'd missed once our few semi-disastrous days of filming were over. At the end, they were going to premiere the video itself. I ran and got Moose so we could watch it together.

David LaChapelle himself played host, commenting on the various scenes and how they were filmed. He was affable and smart, and the costumes and sets looked great. The choreography was spot-on, and, to my delight, the video followed the *Flashdance* scenario closely. Who can forget that memorable scene at the end of the movie, where the young dancer blows the minds of the stuffy dance academy instructors? J.Lo's reenactment of it was amazing!

Then . . . David looked right at the camera—right at me and Moose in our living room—and said, "And then there was this dog . . ."

I let out a little scream and Moose began to lick my face and dance around in celebration. "No, you idiot!" I growled. "This isn't going to be good!" And there I was on MTV, ass in the air, pushing down on Moose's massive backside to try and get him to "Sit!" They had filmed the whole behind-the-scenes debacle, and now the world was watching Moose make a fool of me.

"You see, they brought in this highly trained dog," LaChapelle deadpanned over images of the Tia and Moose Comedy Hour. "But we're still not quite sure what he was *trained to do.*" Well, I guess you haven't lived until you've been publicly humiliated on national television. After years of working with Duke and basking in his reflected glory, this was quite the ego bruiser. But, shaken as I was, I couldn't help laughing when I looked over at my doofus of a dog, J.Lo's beloved Moosey, who had rolled onto his back, stuck all four legs up in the air, and was halfway into nap mode.

★ ★ ★

As the years went on, Moose would prove himself quite the star in his own right. Yes, he had his limits and was mainly cast for his looks—but nobody ever said that supermodels had to be brain surgeons. He stayed true to himself, did what he liked, and refused to take life seriously—which was exactly what made him special. No doubt he changed some minds about "vicious Pit Bulls," but he wasn't the crusader I was. Moose was just along for the ride, and he enjoyed every minute of it. And no matter how bad a day I'd had, he would always find a way to make me laugh.

My big lug eventually died of old age, leaving this world just as peacefully as he had come into my life. And in spite of his shortcomings—or maybe because of them—he left the world a sweeter, happier place.

5
JUNKYARD JOE
From Delinquent to Drug Detector

As I approached adulthood, the menagerie that had kept me in line and out of trouble began to take a backseat to life on the street. More than once during those years, I found myself involved in mischief that bordered on the criminal.

When I was sixteen, a few of my cowboy friends jumped into a delivery truck and took it for a joy ride, only to send it over a cliff in the Santa Susana Knolls. At the time, it seemed like a scene from *American Graffiti*, but looking back, I guess it was more like *Grand Theft Auto*.

Small-town mayhem gave way to something darker as I entered my twenties and moved into the housing projects of Pacoima, California, where I got a taste of L.A.'s infamous gang life. This was "bad company" of a different magnitude. The parties never ended until everyone passed out and the crackle and pop of drive-by shootings was just part of the local soundtrack. One of these almost took the life of Tania's father.

I saw and heard a lot in those days, but something—I'm not sure what—kept me from crossing the line into hardcore stuff. I may have looked like a badass chick, but I never committed a crime and wasn't interested in drinking or drugs. Hell, I never even smoked cigarettes. I did like to fight but I never started one. In fact, what usually got me into the mix was my irresistible impulse to stick up for the underdog. If you were getting your butt kicked for no reason, or were outnumbered and outgunned, you could count on me to jump in and even the odds. Let's face it: This badass chick was always a Girl Scout in a different outfit, just waiting for her real mission to emerge.

Ultimately, the do-gooder in me drew me to the other side of the law. With the idea of becoming a cop or working at a juvenile detention center, I dabbled in some justice classes and went to introductory police academy workshops sponsored by the LAPD. The minute I turned twenty-one, I applied for the police department and was accepted! Just a few weeks later, after I'd already begun training at the academy, that dream died on the vine. I got an official letter stating that my acceptance was rescinded on the basis of a full background check that revealed my flagrant "disrespect for the law."

What the hell? I went right to the board to ask what they meant by this and got an earful. The fact that I'd never committed any sort of crime wasn't good enough for them; gang members had been my friends of choice for a brief period. The verdict was guilt by association. I'd have to find another path.

Running a dog rescue can be tough enough. But running one that focuses on the most maligned breed of dog in the world is even tougher. Thankfully I've always had a creative mind and have constantly looked for ways to profile our dogs in a positive light. With society kicking us in the teeth every chance it got, I needed to find ways to keep smiling and showing that sometimes the old "ya can't judge a book by its cover" was all too true. If anyone should know that, it would be me.

★ ★ ★

In 1997, a one-page story in *People* magazine sent me on a mission that would change a few lives. It was the story of a Pit Bull named Popsicle, who had been found in a black plastic garbage bag inside an old freezer during a drug bust in Buffalo. He was almost dead from wounds he'd sustained in a dogfight, along with starvation and dehydration. With tender care, he came back from his injuries—at which point the U.S. Customs Service decided to help him by giving him a job. He'd be trained as a narcotics detection dog.

To everyone's surprise, Popsicle graduated at the top of his class and went on to assist in the biggest drug bust on the Texas/Mexico border to date. In a raid in Hidalgo, agents seized some 3,075 pounds of cocaine, thanks in part to his nose for drugs.

Popsicle's story got me thinking: Here was a dog who came from a rough and questionable background, just as I did, but luck was on his side. Someone chose to look beyond his demeanor and desperate circumstances and give him a second chance. How could I harness the power of this story to help other dogs? As I was tossing this around in my mind, a call came in from an East L.A. shelter about a "junkyard dog" named Joe.

Joe was a regular visitor at the shelter—he'd ended up there more than twenty times—but his owner always came back to claim him. Finally, the people at the shelter decided that this home might not be the safest place for Joe, so they called me to come in and meet him. Always up for a challenge, I was in my van a few minutes later.

Any dog who had escaped that many times and survived on the street must be a real gangsta, I reasoned. East Los Angeles can be a dangerous place for a human, let alone a defenseless dog. I prepared myself for a scrappy handful of a pooch and listened quietly as a shelter worker told me his back story.

They'd first picked him up from a junkyard, where he'd stood out among hundreds of other strays who came around because he liked to walk along the top of the brick wall surrounding the place. The workers there were intrigued by his calm determination as he circled the perimeter like a soldier on guard duty. When animal control showed up, he simply jumped down and trotted over to their truck as if he had been waiting for them. Joe's owner came to fetch him later that day, full of apologies and excuses.

The shelter's next encounter with Junkyard Joe came as a result of a frantic call from a woman reporting a "vicious dog" keeping her from getting into her car. When animal control showed up, instead of a snarling beast, they found their ol' friend Joe, lounging in the backseat as if waiting for his chauffeur. Apparently he had escaped from his yard yet again and, seeking some shade and a comfy napping spot, jumped through the open window of the nearest Honda Civic. There he lay drowsing until the animal control officer leaned in and said, "Hey there, Joe . . . Nice to see you again, buddy." At that, the dog sat up, hopped back out the window, and trotted over to the animal control vehicle.

These incidents went on for months, and each time, Joe's owner would come and bail him out, swearing to secure his yard better. Each time, the shelter workers would shake their heads and wonder where the Houdini of dogs would show up next. Then one day, Joe went a little too far and caused a scene at a local pet supply store.

The call came in around ten o'clock in the morning about a stray Pit Bull who refused to leave a local shop. Figuring this was a typical "Joe sighting," the animal control officers took their time getting to the store, expecting to find him sitting out front or Dumpster diving in the alley. As they pulled into the parking lot, they saw no sign of Joe, so they got out of their truck and went into the store.

"He's over there," said one of the salespeople, pointing down a long aisle. Sure enough, there was Junkyard Joe, sprawled on the floor of the toy section, nearly obscured by an ocean of tennis balls. As they watched him, he grabbed one in his mouth, tossed it in the air, and, in an attempt to catch it, sent the rest scattering in all directions. Joe had simply barged in through the automatic doors and patrolled the aisles until he found his favorite playthings. After this escapade, animal control had advised the shelter that Joe shouldn't be released to his careless owner yet again—so here we were.

As you can imagine, I couldn't wait to meet the adventurous pooch—but, as is often the case, my expectations were completely confounded. Instead of the giant handful I'd been hearing about, what lay quietly before me was a compact black-and-white fellow—a Pit Bull, yes, but a simple-looking guy, very nondescript. He looked like he should be sleeping on someone's porch.

"Are you kidding me?" I said. "This is the infamous escape artist?" With laughs all around, I agreed to take the little troublemaker with me.

Back at Villalobos, Joe was well behaved and got along with most of the other dogs. He was extremely friendly with humans too, and he had the cutest little prance in his walk—especially when he had a tennis ball in his mouth. Turns out the tennis ball thing was an obsession: He took at least one of them wherever he went. When he ate, he would drop his precious toys around the perimeter of his bowl so as to keep an eye on them. And when he drank water, he would drop one right into the bucket, take a drink, then plunge his head in up to his ears to retrieve his prize.

Joe and I became great companions. We played fetch for hours, stopping only when my arm tired out; he never did. His looks may have been ordinary, but his personality made him stand out in any group of dogs. He was mostly black with a thin white stripe down the middle of his face, and his most prominent feature was his ears. When he stood still, they looked normal. But when he was on the move, they flopped up and down like butterfly wings and never failed to elicit an *Aww . . . he's so friggin' cute!*

Then there was his high-pitched, squeaky bark, which he used primarily when he wanted someone to throw the ball for him. That sound could make you jump straight up in the air if you didn't expect it. His quirks are what endeared Joe to me. It was just a matter of time before I liberated him from the kennel and brought him home to the trailer I shared with Duke.

Duke took to Joe like a long-lost brother—there was real chemistry between them from the start. Within minutes of Joe's arrival, they were racing around my thirty-foot house-and-office-on-wheels, bouncing from the couch to the bed and back again. Space was limited, to say the least, but we made it work. Life as a house dog seemed to suit happy-go-lucky Joe just fine, and he put his Houdini days behind him; he never tried to run away again.

Agua Dulce is a small town in the high desert of the Antelope/Santa Clarita Valley (depending on who you ask). I guess you could say it caters to a middle- to upper-class crowd, though a select few feel they are of an even higher status than that. These particular people have more skeletons in their own closets than probably the entire valley, yet they love to gossip and spread horrible rumors about others. This is especially bad among the women who spend their time sleeping with other women's husbands and shoving cocaine up their noses while judging everyone else's behavior.

I was a loner in this town. Although I had lived there for more than fifteen years, I still didn't feel like a part of the community. Those of us at Villalobos were our own little "country" up there at the end of Anthony Road, but living in a small town and rescuing Pit Bulls was no easy task. Somehow we continued to survive there despite the cold shoulder we got from some of our neighbors.

On school days, I'd drop my kids off at two different schools, then run whatever errands I had to do in town, often stopping at the overpriced local small-town grocery store to pick up a few odds and ends. I was rarely gone for more than twenty minutes. One typical Monday, I pulled up in front of the food store in my van with the customized TUF MUT plates to find a few of the Desperate Housewives of Agua Dulce sitting at a pretty wrought-iron table out front, sipping gourmet coffees and chatting. I got the usual eye rolls and whispers as I walked by in my old denim overalls— whatever, ladies—and quickly grabbed my daily twelve-pack of Coca-Cola. Honestly, I couldn't wait to hightail it back up to the ranch, where the only banshee howls emanated from the coyotes who roamed our hills.

As I walked up to my trailer, I couldn't help but notice something strange about the big front windows: The curtains were gone. How weird. But before I could sort that out, I encountered another little glitch. The sliding-glass door was jammed. After a couple of shoves, it finally gave way, but the whole thing was very odd, as I had never had any problems with it before. As I stepped inside, I tripped over something. My stuff was strewn all over the floor and something had gotten wedged in the door track—which explained that particular problem, but, what the *hell?*

My trailer looked as though a bomb had gone off in it. My four-drawer file cabinet was upended, my computer shoved off the desk onto the floor. Paperwork was strewn everywhere, and sitting in the middle of the mess was Joe, grinning ear to floppy ear. I attempted a scream but nothing came out. My mind was racing, trying to assess the magnitude of the disaster and piece together what happened. Then I noticed Duke. He was lounging on the bed, and I didn't need a Pit Bull translator to help me decipher the meaning of his look: "Yup . . . *he* did it."

My attention turned back to my little black atomic bomb.

"Joe! *What did you do?!*" I shrieked, and in response he began to yip and yap and do his happy dance, completely oblivious to my rage. Each time he spun around, he sent another important file flying across the room like a Frisbee. Just when I was about to lunge at him, he stood stock-still and trained his attention on a spot near the ceiling. He was staring at a shelf up above where my cabinets used to stand, before he'd overturned them like bowling pins. He started barking frantically like Lassie when his best buddy fell down a well.

"What the hell is it, you crazy dog? *What?*" I shouted, but Joe just kept bouncing up toward the ceiling as if he wanted to sprout wings and fly.

Then I saw it. I knew what my persistent little Pit Bull wanted. It was a tennis ball.

If you're a dog person, I'm sure you can relate to this particular recipe for disaster. We all keep dog stuff around: treats in the glove compartment, a leash in the purse, a tennis ball up on a shelf next to the stapler and extra office supplies. Okay, well that one was pretty dumb. I swear, I don't even remember putting it there—but apparently, I had. I put a bright neon-green ball of temptation on a high shelf where Joe could see it but couldn't reach it.

I found a stepladder and climbed up toward the treasure. As soon as I retrieved it, Joe got even crazier, literally bouncing off the walls and any furniture that had miraculously remained standing. I stared at the ball in my hand for a second, not quite believing that this fuzzy little object could be the cause of such devastation. Then I tossed it into the air and, with a graceful leap, Joe caught it dead center in his mouth. He pranced off down the hallway, plopped himself on the floor, and began suckling the ball as if it were a pacifier. What could I do but begin the process of reassembling my life?

As careful as I tried to be, these tennis ball incidents were all too common. If a random object even remotely resembled a ball, Joe would take out everything in his path to get to it. I became convinced that he was suffering from a full-blown obsessive-compulsive disorder. Sometimes he would train his steely glare on the couch and bark nonstop. Thinking that he had lost his mind, I would attempt to show him that he was barking at *nothing*—only to discover upon investigation that there was in fact a tennis ball or other toy hidden deep within the couch cushions.

Joe's addiction became such a problem that I couldn't leave him alone for even a little while, for fear of what mayhem I'd find when I returned. In his world, there was a tennis ball around every corner, inside every wall, behind every piece of furniture, and at the bottom of every trash can—and he couldn't rest until he'd ferreted it out.

As a dog trainer, I had been taught that it is easier to turn negative behavior into positive behavior than to suppress it completely. How might this apply to my ball-crazy dog? I was spending a lot of time trying to solve this riddle when I met some phenomenal dogs that provided a clue as to how I might retrain Junkyard Joe.

By this time, Villalobos had gained some renown in the dog world, and it was common for us to be invited to various dog functions. One of the more interesting ones—and a first for us—was a community-awareness event in which police dogs from local K-9 units would compete against one another to show off the tasks they'd been trained to perform. Some had been trained to apprehend "bad guys," others to find lost children, still others to sniff out drugs or explosives. Some were even termite sniffers! These dogs were nothing short of amazing, and I was beyond intrigued.

When the demonstrations were over, the public was encouraged to walk around and meet the "super K-9s" and their handlers. My awe and curiosity were overflowing as I interrogated one police officer after another, asking them a million questions about their training techniques. I was especially interested in the detection dogs: those trained to sniff out drugs and bombs. As I suspected, they'd all been trained using toys. It was essential that they have an insane toy drive!

Now, who did that remind me of?

Although I'd never gotten to pursue my dream of a career in law enforcement, getting involved "dog wise" might give me the feeling of helping; I'd be living the dream vicariously through Joe. I learned that I didn't have to be a cop to have a dog trained in detection—so I began to do some snooping.

I knew Joe had what it took. He'd search for a stray tennis ball to the point of exhaustion, then get up and do it again. I was thrilled at the thought of putting this compulsive behavior to good use, so I began calling local police departments, looking for someone with the skills to help me turn Joe into a drug dog. In several instances, I was told that the training programs were limited to law enforcement personnel—but a few kind people took the time to steer me to outside trainers, and I refocused my search.

The trainers I talked to were receptive and interested in my story—until they heard that Joe was a Pit Bull. Then they couldn't get off the phone fast enough. But somewhere along the way, I had piqued the interest of two fellow Pit Bull owners who wanted to join Joe and me on our quest. Presumably, Goose and Chopper had the "right stuff" too. All we wanted was to enjoy the experience of motivating our dogs to do good work while having their favorite form of fun—but we were having trouble making headway against the tired old prejudices against Pits.

Just as I was ready to give up on the whole enterprise, the phone rang. The trainer for the Ventura County Sheriff's Department finally returned my call and said that some of the deputies in his department were interested in trying out a Pit Bull as a drug dog. To be honest, his proposition sounded more like a challenge than an invitation: Prove to us that your dogs can do this kind of work. My fellow trainers and I were all too eager to pick up that gauntlet.

On the first day of "school," we were told that the dogs would have to pass a test to determine just how "ball crazy" they really were. The instructor had us take them out to a field and play ball with them for a few minutes while he observed. Mind you, it was pretty hot outside that day; it wouldn't take long for the average dog to poop out and look for shade. Not our dogs. After a good ten minutes or so, when the pups showed no sign of slowing down, the instructor set a water bucket out in the middle of the field, near where we were running the dogs. Would they stop playing fetch in order to take a cool drink, or would their attention remain riveted to the task of retrieving balls from the high grass?

When it was Joe's turn, we went through the routine and played ball for quite some time. And just as Joe's tongue was dragging on the ground, the instructor brought out the water bucket. Standing there in the summer heat, I wanted to drink out of it myself, and wouldn't have blamed Joe for doing so—but this was the "Joe Show" and it had to be determined . . . how determined he was. It was all about *drive*.

As my little lump of coal sped into the tall grass to look for the tennis ball, all I could see was the flag of his tail sticking straight up in the air. Then he came bolting out, prize in his mouth, and began racing back to us. As he neared the water bucket, I have to admit, I held my breath. I saw Joe give it the eye and slow down slightly, but he gave that bucket only a fleeting thought, then kicked it into high gear and raced right to me. As he slid up to my legs like a cutting horse trying to thin out the herd, the dirt flew; Joe flung the ball at my feet and let out a bark that could only be translated as, "WHOO HOO, I DID IT, I DID IT!" And there he sat, tail wagging so hard that he was making dirt angels.

It was then that the instructor gave me the okay for Joe to go and get some water. As I motioned for my overly proud boy to take a break—as if passing the test wasn't enough—he once again grabbed the tennis ball, ran with it, dropped it in the water bucket, and did his signature drinking trick: He submerged his entire head past his eyes and kept a close watch on his prized possession.

Junkyard Joe had passed his test with more than flying colors. From Dumpster diving to drug detection, this little Pit Bull was about to leave a life of crime to help prevent it.

Our training would take only two or three weeks, which was a pleasant surprise to me. You'd think that teaching a dog to detect four different types of drugs (marijuana, cocaine, heroin, and meth) would take much longer. But that just goes to show you how much smarter dogs are than humans.

The first lesson I learned was that although this was to

be a very positive experience for the dogs, they would not be receiving any type of food or treat as a reward. Their love and obsession for their toy would have to be all they would need and want. So the first step was to teach them to "Find it!" combining a verbal command with the actual act of looking for their toy. By this point, we had switched our dogs from tennis balls to Kong toys, so that eventually we could stuff scented cotton balls (substituting for actual drugs) into the toy. That's how they'd learn to associate one with the other.

Our first training field was an old abandoned house, trashy and unkempt, similar to some of the drug dens the dogs might encounter on the job. The extra obstacles strewn about were meant to challenge the dogs as they looked for the toy.

With a toss of the Kong toy and the command to "Find it!" away my little man went, catching on to what was expected of him right away. His two fellow classmates were equally quick studies.

As we each tossed our Kongs into the house and gave the "Find it!" command, Goose and Chopper methodically searched the house. Slowly analyzing each square inch of each room, they eventually found their toys, then "sat pretty" and looked toward their handlers. Their moves were precise, controlled, and perfect.

Then there was Joe.

After just a couple of preliminary tosses and searches, my little pistol was ready to rock 'n' roll. Before I could even shout "Find it!" he was barking frantically. He knew what he was there to do and he wanted to get to it! Joe's searches were turbocharged rather than methodical: He'd fly over the top of a couch, overturn tables, burrow under rugs, basically bulldoze the room until he'd found his prize.

Joe was over-the-top excited about this new school for which he had signed up. It was giving him an outlet for his nonstop OCD behavior. But when it came time to teach the dogs to associate the drugs with the toys, my first thought was: *Now we're going to add* drugs *to this equation?* The reality was nothing like I thought it would be. Counter to the rumors that I had been hearing for years, you don't get the dogs hooked on drugs. Joe wasn't the only one learning something new: I was about to get schooled myself.

Because our instructor worked for the sheriff's department, he was legally allowed to handle drugs for training purposes. When he told me that Joe would be trained to detect marijuana, cocaine, meth, and heroin, all within the brief training period, I was dumbfounded. How could he learn all of this so quickly? As if reading my mind, Joe locked eyes with me, whipped his tail back and forth, and let out a high-pitched bark that said he was more than up to the next challenge.

The training concept was just an extension of what Joe had already learned. Drugs (wrapped securely in plastic) were inserted into the Kong toy and it was then tossed into a room of the house. Retrieving the toy was once again a reward in itself—but now the scent of the drug was becoming associated with the toy. Eventually, the Kong would be eliminated and just the drugs hidden throughout the house. As soon as Joe found the contraband and "alerted," he would be rewarded with his toy. And that's when Joe hit a little speed bump on his way to the head of the class.

As each search was conducted, Joe showed over and over again that an "aggressive" alert was much more fun than one of those boring "passive" alerts. He'd bounce up and down, barking frantically and spinning in circles. Each search excited him more than the last. This dog from East Los Angeles had finally found his niche, and he was determined to perform over and above the call of duty.

On one particularly warm and breezy afternoon, Joe got a good whiff of his new job. We were working on detecting marijuana. Our instructor had decided to up the stakes a bit and extend our search to the second story of the house. I gave Joe the "Find it!" command and away he went, bouncing off the walls like Sonic the Hedgehog. I attempted to step in to slow his roll, but the instructor motioned for me to stop and let him carry out the search in his own way. *Okay,* I thought, laughing at my four-legged demolition crew. Once he'd scoured each downstairs room, Joe raced up the stairs to a bedroom. Following behind, I motioned for him to search the room, but the second he was through the door, he stopped dead in his tracks, put his nose up in the air, and froze, trying to catch the scent. Then, before I could blink, he raced for the open window and reared back as if to jump out. The instructor and I both leapt at Joe at the same time, and between the two of us we managed to grab just enough of him to prevent the overly zealous drug hound from taking the plunge. Once I had a firm grip on his collar, I looked up at the instructor. *What the hell was* that *about?*

He explained that the breezy weather was the culprit: Joe must have picked up the scent of drugs down below, through the second-story window. This confused me, because we'd scoured the entire downstairs. What did we miss? As his handler, I was supposed to help him; if I felt he was overlooking an area, it was my job to point it out and encourage him to keep looking—so the oversight was as much mine as his. We went back downstairs as I tried to banish the mental image of Joe leaping to his death because of my own carelessness.

When we got to the bottom of the staircase, Joe and I noticed something at the same time, though I used my eyes and he his nose. There was a small door we had missed. Joe began barking and spinning in circles as I cracked it open. He shoved himself in nose-first and began working what turned out to be a laundry room. Within seconds, he alerted by the washing machine, and before I could reward him, he'd jumped onto—and into—the top loader head-first, his muscle butt and back legs sticking straight upward, his happy tail whipping back and forth like a CB antenna as he dug for his prize. Although a bit overeager at times, Joe was on his way to being a kick-ass drug detection dog.

★ ★ ★

After two weeks of training—in a feat that was nothing short of amazing to me—Joe and the other dogs had learned to detect the scents of all four drugs flawlessly. But before they could graduate, the three would have to prove themselves in more realistic, high-pressure environments. They'd have to track down the drugs in outdoor buildings, scrap piles, and vehicles.

One of Joe's final tests was to search through an outdoor pile of junk and debris for meth. I don't mind telling you, I was a little nervous. Besides wanting Joe to succeed, I had a lot at stake as well. I'd been angry when my own chance at a law enforcement career had been yanked away, and that anger had festered inside me for years. Now, with the help of my rehabilitated canine criminal, I was going to prove that underdogs can excel brilliantly when given a second chance. Not only that, but Pit Bulls weren't traditionally permitted to do law enforcement work; that honor was typically reserved for German Shepherds or Belgian Malinois. We were groundbreakers, and I was proud that this work might help pave the way for other Pits to be deployed and respected as never before.

There was a slight breeze, and I could see Joe was getting confused as he raced back and forth across a stack of plywood, not sure quite where the drugs were. I took a deep breath and got permission from our instructor to regroup for a minute, since we both had a lot riding on this. I calmed my four-legged firecracker as best I could, then reiterated the "Find it!" command. After a couple of swoops back and forth, Joe finally alerted by wedging the front half of his body so far under a piece of plywood that I had to pull him out bodily. Score one for Joe, zero for the meth.

The other dogs were performing equally well, and finally, we were just one test away from being the first all–Pit Bull drug detection group. Joe's last challenge would be to find some marijuana during a vehicle search. If he could do it, he'd be certified as a legit drug dog. Word of the team's final exams had gotten around, and some of the local K-9 handlers had shown up to watch. As I waited for our turn, I overheard one of them say, "Let's see what these Pit Bulls are all about." Fired up, I bent over and kissed Joe right on top of his unsuspecting head, and before I could pull away he laid a sloppy one right on my lips. I might have been a little nervous, but he was cool as a cucumber and ready to prove to the world that Pit Bulls can do anything they set their minds to.

We'd been taught that to do a proper vehicle search, we had to make a quick perimeter sweep around the outside of the car first. Then we were to direct our dog to go inside the car. After a once-over of the entire vehicle, we could go back for a more precise search. But because Joe got so amped up during his searches, I decided we should do our own thing and head straight for the interior of the old beat-up undercover police car.

I reached to open the door, and in typical Joe fashion, he jumped up and flew through the open window. Like a raccoon rummaging through a trash can, Joe began to destroy the inside of that car. When the marijuana could not be found, he decided to move his search outside and, like Superman, flew out the window and looked up at me as if to say, "Let's do this!"

Our perimeter search took us around the bottom of the car and over the top of the tires. Then, as we neared the exhaust pipe, Joe exploded in a ball of fire. He put his front end to the ground in what looked like a bowing position and barked frantically. Just as I went to tell him "good boy" and reward him with the toss of his toy, Joe's excitement got the best of him and he grabbed the exhaust pipe in his mouth—and attempted to yank it off of the car.

"DON'T LET HIM . . . " yelled our instructor, but it was too late. Like the snapping of a turkey wishbone, the pipe came off with a *crack* and Joe began to fling it around. The wrapped bag of marijuana fell out the end of the pipe, and the instructor and Joe went for it at the same time, causing a near miss of that sack of weed being strewn about the dirt. As Joe barked that bark that said he was very proud of himself, I tossed his Kong up into the air, and, on the first attempt, he caught it airborne.

As a result of that crazy scene—which was like something from a Cheech and Chong movie—Joe graduated.

Still disappointed about being denied a career in law enforcement, I began to live vicariously through Joe. Not only did I feature him in my educational seminars, I offered our services to facilities that needed to be swept for drugs. Witnessing the awe on people's faces as they watched my street dog—my Pit Bull—doing an important job felt pretty damn good, I must admit. I'd done what even I had feared was impossible: I'd turned Joe's negative behavior into something positive. And in the process, I'd broadened people's minds.

So many "problem dogs" are dumped in shelters or abandoned on the streets every year when a little bit of training and some extra attention might turn them into great companions. I understand that every dog can't work in drug detection, but I still think Joe's story is instructive. Think about it: It took just two weeks to train this nutty, highly destructive dog to do something useful—and he enjoyed every minute of it. It gave him a purpose and something to which to look forward. At seminars and on service calls, Joe's optimistic attitude and high-pitched "yip" became his trademarks. Through his work, Joe became a genuine superhero—and he even stopped tearing up the house. Well . . . mostly.

By the late summer of 2007, Joe had been making the rounds for about two years and had become a welcome presence at schools, community centers, and juvenile halls around L.A. We even got occasional calls from parents who wanted to hire him to search their kids' rooms for drugs!

At that point, my relationship with Mariah's dad, Jon, had finally ended and my days of sleeping in the small travel trailer were over. When Jon moved out, Joe, Duke, and I joined my children in the 4,000-square-foot southwestern-style house, where we had a lot more breathing room. Each morning, I'd wake up early to get the kids off to school—but before my feet even hit the floor, Joe would spring off the bed and slither underneath it army-style, to search for his tennis ball. He'd appear a few seconds later with his prize, barking as best he could with his mouth full until I told him to "Drop it!" so I could throw it down the long hallway. Duke, meanwhile, preferred to lounge in bed a little longer, clearly feeling that such behavior was beneath his dignity.

One morning I woke up around six o'clock as usual, to find no sign of Joe. He'd been there when we went to sleep, but now only Duke lay next to me. I called out to him. No response. Then something else odd happened. As I stepped out of bed, Duke got up too, only to lay down on the floor with his head on his front paws, eyes trained under the bed.

A wave of nausea washed over me and my heart started to pound. I knew something was very wrong.

I bent down to Duke's eye level and there was Joe, sprawled out under the bed. As I called gently to him, his eyes turned in my direction but he didn't move a muscle. Carefully, I reached under and pulled him out as Duke stood by protectively. Joe's body was limp, and he groaned as I stood up with him cradled in my arms.

Yelling to Tania that she was going to have to stay home from school and babysit the rest of the clan, I placed Joe on the passenger seat of my van and raced down the long dirt road to the main highway. It was nearly an hour's drive to the vet, and Joe remained quiet the entire way.

I'd been with the team at East Valley Veterinary Clinic for more than fifteen years and I would have trusted them with my own life. I knew they'd do what they could for Joe. They checked him in and I sat in the waiting room while they ran all sorts of tests. It would take a day to get the results, so, late that afternoon, I kissed my alert but immobile pooch on his sweet nose and made what seemed like the even longer trip back to Agua Dulce.

Night was falling fast by the time I got home, and I was greeted by the yowling of the fifteen coyotes who lived freely on our property. Soon the chorus was joined by the wolves and wolf hybrids who still lived there as well. As I sat back in my seat, watching the desert dust catch up to my rolling tires, the wild animals frolicked up and down the hills, waiting for me to come up and feed them. I stopped my vehicle and sat longer than usual, watching the sun get swallowed completely by the horizon. To my ears, it sounded as if the coyotes were as sad as I was.

This was all before the days of our TV show, when we were still making the rounds of the networks, pitching the idea for a series we wanted to call *The Underdawgz*. The day after Joe went to the hospital, we had a meeting scheduled with some bigwigs at Fox and I had no choice but to attend, though I knew I'd be an emotional wreck.

I spoke with the vets first thing in the morning, before I left the house, but they still had no idea what was wrong with Joe. His condition hadn't changed—he still didn't seem able to move and was neither eating nor drinking.

I managed to smile occasionally during the meeting, nod as if I were paying attention, and mumble a few remarks on cue—though my mind was on Joe. Whether I was discernibly off my game or the network suits just didn't share our vision, it was clear they had no interest in our show. I honestly didn't care; I just wanted to get out of there and go see my poor sick friend. As I walked through the big glass doors at the network, my phone rang.

Dr. Hohne had always been the epitome of calm professionalism, but this time was different. His voice cracked, and I could tell he was having difficulty finding words.

"Hey, Tia . . . I am . . . so sorry. I'm just so sorry. Joe didn't make it. We're not even sure what . . ."

At that, I just plopped myself right down on the curb, my kneecaps just inches from a stream of speeding West Los Angeles traffic. I sat there in numb disbelief for a few . . . minutes? Hours? When I could finally trust myself to move, I got up, found my van, and made my way home.

The moon that night was full, with dark, wispy clouds swirling across the face of it. As I stepped out of the van, I heard a young coyote let out a couple of "yip yip"s and then stop.

Then it came: a full-blown chorus of coyotes, wolves, wolf hybrids . . . even Pit Bulls. There were howls of every pitch, tone, and rhythm, rising and falling and spreading out across the desert. It was as if they knew they'd lost a brother.

Our big house felt empty. No Joe barreling down the hallway, plowing through a herd of cats who never even bothered to get out of the way anymore. LuLu, our tiny, feisty Brussels Griffon mix, stood near the entry and cocked her head at me, wondering where her big friend was.

As my steps echoed across the Spanish adobe tile on my way to the bedroom, I couldn't help but think back to the beginning of my journey with Joe. He had survived the mean streets of East Los Angeles without so much as a scratch, wreaked his share of havoc, and stepped up to become a genuine role model. Now I had to deal with his departure—so sudden and so completely inexplicable. My grief gave way to anger as I thought about all the good work he still had to do.

I plopped down on my bed, face first, but overshot it a little, so that my head was hanging over the edge. Then my eyes focused on something green. A tennis ball.

Still hanging upside down, I stuck my face farther under the bed and saw an amazing sight: wall-to-wall neon green balls—dozens of them! Where the hell had Joe gotten them all? I knew it was his hiding place, but I had no idea Joe was so "rich."

I couldn't help but smile. Dogs are such special beings. They feel our every emotion. They hear, see, and smell things that we don't even know are there. Perhaps they sense their own impending mortality as well. I had to think that Joe had begun hoarding his treasures for a reason. Maybe he hoped to take his cache of playthings on his final journey, so he could play fetch throughout eternity.

I couldn't help it. I picked up one of the tennis balls and tossed it out the back door and into the darkness of the desert. Then I smiled. Somewhere out there, in another lifetime, I just knew that Junkyard Joe had nailed that catch.

6

MONSTER

A Dog and His Boy

As children, we grow up learning to fear monsters. There are the ones that lie in wait under our beds, keeping us frozen in terror as our legs dangle over the sides. Then of course there are those Sasquatch-size creatures who hide in the closet waiting for us to fall into a slumber, only to leap from the clutter of clothing and mismatched shoes, leaving us hyperventilating and trying to call out to our parents. Even if our parents manage to comfort us, we may never get past that fear, as our imaginary, disfigured villains become all too vivid. Eventually the bulging green eyes and dripping fangs become baby blues and veneers as pretend turns into the real-life monsters: humans. If only I had known there would be many monsters in my future just like that . . . Hiding under the covers may have been the safest place my kids and I have ever known.

These were some pretty dark times for our family. The rescue was at its capacity, and the money and donations had all but dried up. And despite not being the most religious people in town, we found ourselves having to beg for food from the local church. It's amazing how much hunger can play a role in making one quite the hypocrite. I was trying to hold it together for not only my two-legged kids but the four-legged ones as well. And despite these desperate times, I continued to follow through with my mission of rescuing dogs.

By now, Villalobos Rescue Center had a solid relationship with all the local animal shelters. Although they all received hundreds of dogs on a weekly basis, every now and then there would be that "special" one who would tug on the emotions of one of the shelter workers. Such was the case on this day.

I was told that a dog was coming in and that animal control wanted me to bring him to the rescue rather than have him entered into the shelter system. So I jumped in my beat-up white Dodge van and made the one-hour drive into the San Fernando Valley and to the East Valley Shelter.

As I walked through those double glass doors, as I had so many times before, the scene was unlike any I had ever encountered. Right away I recognized the animal control officer (ACO) with whom I dealt on a regular basis. She was talking to a family with three children, one of them very distraught. As I entered the lobby, the ACO showed an obvious look of relief. Realizing that this was why I'd been called down to the shelter, I walked over to the family and introduced myself.

They appeared to be from a nice, traditional family that, like so many others, had fallen on hard times and had to move. They had tried desperately to find a place that would allow them to take Monster, the Pit Bull of their eleven-year-old son, but were rejected at every turn.

Looking down at the extremely large black, gray, and white splotched (known as blue merle) dog with his pointy, demonic-looking ears, I learned that he and the boy had been together since Monster was a puppy, and now the family was having to make the most heart-wrenching decision to tear away the best friend of their middle son. After trying to help the parents come up with alternatives, the time came for me to do probably the hardest thing I'd ever done.

The father hesitated but was finally able to free his son's grasp on Monster's leash. As I took it from him, the son burst into uncontrollable crying and sobbing. Monster knew something was wrong and began to lick the boy's face in an attempt to wash away his gushing tears. The youngster then flung his arms around the neck of the big dog. As his mother slowly tried to pull her son away, it became an emotional game of tug-of-war. I stood there feeling a wave of despair for this child. As the boy's crying turned into banshee-like screaming, only then was I able to finally free Monster from his grasp.

The family members embraced each other and made their way out the door. With tears streaming down her cheeks, the mom mouthed the words "Thank you" before the glass doors slowly closed. Monster began to squeak and whine, his unbreakable stare fixed on the reflection of his boy in the glass door. Even after the boy was long gone, the heartbroken dog refused to move. Emotionally drained, I sat on the lobby bench to catch my breath. In all my years of running this thing called a Pit Bull rescue, I'd never had my heart ripped from my chest like that.

I stroked Monster's head, trying to comfort him. The shelter supervisor came out and sat down, and both of us attempted to tell the dog that everything was going to be okay, although neither of us believed it. We talked about how a "job" saving dogs could suck so much on days like this. And with that thought, Monster and I left the shelter, the sound of desperate, barking shelter dogs behind us. If only Monster knew that he'd come within minutes of being one of those inmates in an already overcrowded city shelter system. But that wouldn't have taken away the pain of his broken and confused heart.

We made the long drive home, but I just couldn't bring myself to put this family pet into an outside kennel. It was one thing to find a stray and starving dog in the street and house him or her in that type of environment. Going from the mean streets of Los Angeles to a kennel and a bundle of cushy blankets stuffed in a doghouse—that was a major step up for a street dog, as far as living situations go. But despite his tough name, this dog was definitely not a monster at all.

I set Monster up in the room belonging to one of my twin sons. Because Tania and Mariah already had their own dogs in their rooms and Kanani at the time was housing a cat, Moe's room was the choice by default, as it was the only animal-free room. Monster and I entered what was nothing short of a clothing land mine as I tripped over piles of shoes, work boots, dirty T-shirts, and jeans. Of all my kids, Moe and Mariah were my "dirty" children, and there was many a day when the thought of bringing in a shovel to dig my way through their disasters almost became a reality. But it was apparent that my threats fell upon deaf ears, for here I was, standing knee-deep in a teenage war zone.

I was picking up Moe's muddy motocross gear and throwing it in the already overflowing hamper when Monster's behavior put a sudden lump in my throat. He had found comfort, both physically and emotionally, in a pile of my son's clothing. He had made himself a little nest and was already curled up in a ball, head tucked in under his own paws. At a quick glance, he seemed at peace, but I could hear him groaning slowly and sadly as he submitted to his new life without his boy. I leaned over and kissed him gently on top of the head, hoping that it brought even the slightest feeling of relief to him. I slowly closed the bedroom door and left him to fall asleep in his own sorrow, hoping that the next day had something to offer that could bring him out of his depression.

Later that afternoon, I was outside in the kennels doing the second shift of cleaning when I heard Moe yelling for me from somewhere on our ten-acre ranch. He had come home from school, and he looked bewildered.

"Yes, my favorite son," I said, teasing him sarcastically. He was my child who always got into the most mischief. "What's up?"

"Mom . . . um . . . there's a strange dog in my room. He's got spots all over him!" And with that he made me smile— not only by his simplistic way of describing the blue merle-

colored dog but also by reminding me how my children had
to deal with my unpredictable animal rescue efforts.

But Moe was my most vulnerable. He and his brother
had a tough past, and there were many family issues, some
things too painful to even talk about. When Mariah had
brought them home from school and they needed a place
to live, I accepted them for who they were. Raised by my
nonbiological mother myself, I treated these boys as though
they were my own. Getting a taste of what life would be like
living with a woman who ran a large dog rescue would soon
be a case of "expect the unexpected," and today was one of
those days.

There was the time that a movie animal training facility got shut down and I temporarily took in some of their
animals. The little red fox was so afraid that I put her in
Mariah's room. Then the two very talkative Macaw parrots and five-foot-long (with tail) iguana went into Tania's
room. And let's not forget the (pretty intimidating) badger
who was housed with Kanani. I still remembered the very
surprised—and, okay, shocked—screams from my children
when they came home from school and found their unexpected guests.

"Well, my son, he's a sad dog and he lost his family, so
he's going to stay with us until he gets a new one," I said
to Moe. He just stood there and made this face. Although
my kids loved animals and helped me with the kennels,
he wasn't really into having a dog. He was into his dirt
bike and motocross phase, and that's what he spent his
off time doing.

"Mom . . ."—he dragged the word out, even going into the whining phase—"I don't want a dog in my room."

"I know, but he's sad, so let him stay with you until we can make him happy."

Moe saw things in black and white, so he liked explanations to be given in the same manner. He didn't like long, drawn-out conversations or metaphors or analogies. Just tell him like it is. So the words *sad* and *happy* were all he needed to hear, and he was content.

"Okay, Mom . . . he can stay, and I will share." Just like that, Monster had a new boy to watch over him . . . for the time being.

Weeks turned into months and months into years, and after two years we finally had an adoptive home for Monster. Moe had gotten used to having a dog of his "own," but at the same time, my kids understood that we were in the "business" of finding dogs homes. So with a sad face, Moe stood in the dirt driveway as one of my volunteers pulled out and made the long drive up to northern California on what we hoped would be Monster's final journey.

Monster wouldn't have a boy of his own, as his new owners were a middle-aged couple who had a small ranch. They had a secure and stable lifestyle and a few horses. The wife worked out of the home, so she would be with Monster almost 24/7. It was a dream home for any dog.

Moe, on the other hand, looked lost. When I asked him what was wrong, of course he would answer with "Nuthin'," but I knew. He sat in his room not really doing anything. He was going through some tough teenage things, and I think having "My-Ya" (the nickname he had given Monster) around gave him someone else in whom to confide. Although he and his twin brother, Kanani, were very close, it was Monster who lay in bed with him at night, falling asleep to the sound of the late-night infomercials and playing an occasional video game or two.

At times like these, I wondered if this was the right life for me to drag my kids into. As a single mom, having the four of them working with the rescue, seeing happy times with adoptions and sad times with some of the rescue situations, was my way of teaching them responsibility and compassion for life. But it was also an emotional roller coaster, one that even most adults couldn't handle. Yet here I was, giving them no choice but to join me in my "hobby." And because they were good children, they gave me no grief, no complaints, and they did me proud by helping each day, no matter what degree of heartache came their way.

Moe would eventually get over the initial vacancy he had in his life left by the "spotted dog," but he would ask from time to time how Monster was doing. There is a saying in the rescue world: "No news is good news." Meaning, if an adopter isn't calling us with a complaint or to return the dog, then all is good. But a year later, "the call" came—but not from the adopter.

There's nothing that makes your heart race more than getting a voice-mail message that starts off: "I found a dog with your name tag on it." Every dog that we adopt out leaves with our ID tag on their collar. One of the stipulations in our adoption procedures is that our tag (along with the adopter's) must remain on the dog's collar. The main reason for this is that we monitor our calls almost twenty-four hours a day, so we'll hear about any lost VRC animal more quickly than the average person.

Once I had the Good Samaritan on the phone, I began to question her about the identity of the dog and the situation in which he or she had been found. She explained that she'd found this particular dog rummaging through trash down the street from her house. She said he was specifically trying to get the last nibbles out of an old can of cat food. The dog was pretty skinny and very calm, with no emotion. Then she said the word "spotted." I let out a huge gasp, knowing there was only one dog who matched that description. It was Monster.

I asked the Good Samaritan where she had found Monster, and together we realized that it was in the same neighborhood where Monster's adopters lived. My first reaction was to be angry because he was found skinny, but the adopters just didn't seem like the type to starve a dog. They had been financially well off, so it wasn't as if they couldn't afford to feed him. And we like to pride ourselves in being thorough when it comes to adopting our dogs out.

I asked the woman if she could take Monster to her house while I called our adopters. I called for hours and got no answer, so I called Monster's rescuer back and informed her that I was at a dead end. She then offered to go over to our adopters' house, and although I didn't want to put more responsibility on her, I was desperate and confused. So I gave her our adopters' address, which turned out to be literally around the corner from her.

Within a few minutes I got a call back—and the story the woman told me was overwhelming to hear. Apparently when she got to the adopters' house, she recognized who lived there—or, shall I say, *used* to live there. She explained to me that the wife had gotten cancer and had quickly succumbed to the disease. The husband had fallen into a deep depression and had basically given up on life. The woman told the story of how Monster's adopter let the beautiful mini-ranch become overgrown with weeds, and there was trash thrown about the front yard. There were no more animals on the property, and the rumor around the neighborhood was that the husband had "disappeared." All we could assume was that Monster had somehow gotten lost in the shuffle, or maybe a relative or friend of the family was supposed to come and get him. We would never come to find out, but now more than ever, a guilty feeling came over me. I had tried so hard to find Monster his happy ending, and although it wasn't my fault, I had failed him.

The next day, the kids all got home from school at the same time. Mariah sat down in front of the TV with her afternoon bowl of cereal while Tania marched straight to her room to talk on the phone with friends. Kanani went out to the garage to tinker with motorcycle stuff, and Moe went to join him.

But I stopped him. "You need to clean that dump of a room," I barked at him. This came as no surprise to Moe, because it had become a daily chore for my "messy" child. Without any resistance, he walked straight down the hall, and I heard his door close. And within seconds, the sound of a very happy boy came bellowing through the door.

"My-Ya!" With a smirk on my face that was soon followed by happy tears, I slowly opened Moe's bedroom door and peeked in. He was on the floor, among several nests of dirty clothes, spooning with the very happy spotted pooch. Between the thumping happy tail, slobbering tongue, and the obviously ecstatic teenager, just like that, Monster had found his new boy.

Moe and Monster became inseparable. They went everywhere together, and Monster became very protective of my son. It got to the point where only the immediate family could get near Moe without fear of being chased down. It was as if Monster knew that Moe needed to be watched over. Moe and Kanani were going through some difficult times with some outside family issues, and things were tough for the boys. The more Moe drifted off into his own world, the more Monster stood watch.

I cannot count the number of times during these dark months that I had to go looking for Moe and Monster when they ran away for the day and off into the desert mountains. As much as I understood why my son needed some alone time to clear his head, I used to worry about him being off by himself. But now, with his trusted four-footed friend at his side, I could breathe a little easier, knowing he would be well taken care of.

I overheard many painful conversations between Moe

and Monster through that bedroom door. That big, scary-looking Pit Bull gave my son some relief and a sense of security—something that I sometimes just couldn't give him. As much as I was the "cool mom," sometimes kids just don't want to open up to their parents. I could see that Moe was in pain, and there were things in his life that were dragging him down. But opening up to Monster was much easier for him, and he leaned on him for all his strength.

Like the time in the French Quarter in New Orleans.

We had already decided to move from California to Louisiana. This was for several reasons, but mainly for two. The first was financial. Life in Los Angeles County was taking its toll on us, and we were on the brink of losing everything. The second reason was to get my kids—mainly the twins—out of the area. The kind of people living there had changed drastically since we'd moved in, the area saturated with drugs and the people becoming less and less friendly, and it was affecting my family. It was time to relocate.

Although we had decided on Louisiana for our new home state, we were still considering exactly where we wanted to land with Villalobos. It took us several trips back and forth across the country, looking at various properties, as we tried to find not only a good environment for our dogs but one that would accept us as well. And with each excursion, we would "camp out" in New Orleans as we explored the rest of the state. And nothing says "New Orleans" like the French Quarter.

Our home base became the Olivier House Hotel. Not only was it unique—the coolest, most haunted place in the

"FQ"—but they also loved dogs, so the sight of multiple kids, tons of luggage, and even more dogs did not scare them off. But it was during one of these visits that the twins, being typical brothers, got into a pretty heated boxing match.

We were all getting ready to leave the hotel to explore the French Quarter, eat some lunch, and then head out to look at some potential new locations for the rescue. As I sat on the edge of the beautiful antique bed in my streetside room, tying up the final lace on my work boots, I heard the distinct sound of a *thump* hit my wall. And then another. Then an even louder one. I recognized that sound all too well, and, having become accustomed to refereeing twin teenage boys, I sprang into action.

By the time my friend Louise and I busted into their room, the sweet aroma of fruit punch filled the air—not only in essence but literally in the physical sense . . . as in all over the walls. Every piece of furniture was tipped over, every article of clothing they owned strewn about the room. I grabbed one twin by the arm and Louise grabbed the other by the shirt. As we tried to end round one, Moe became agitated and called to Monster, who ran down the hall with him. We were about to deal with another "running away" moment.

Somehow my son and his faithful sidekick had managed to disappear into the winding corridors of the old hotel before I could figure out which direction they went, and once out onto streets of the French Quarter, it would take nothing less than a pack of Bloodhounds to find them. At this point, the catalyst for the fight no longer mattered.

My family was still very new to the area, and the rumors of the dangers of Bourbon Street still kept us looking over our shoulders. Although Moe was twenty-three years old, he was still at times a young and naïve "little boy" to me. Louise, the rest of the kids, and I fanned out throughout the Quarter, frantically looking for my angry child and his dog.

After about an hour, my cell phone rang, and I didn't recognize the caller ID. Normally I would never answer this type of call, but my motherly sixth sense told me to pick it up. The voice on the other end was a familiar one.

"Mom . . . I'm lost." It was Moe.

"What? Where are you? Whose phone is this?" I tried not to panic, but my emotional mom voice took over.

"Mom, my friend is letting me use his phone," he replied, matter-of-fact like. "He's going to help me get back to the hotel." And before I could answer him through the din of the loud music in the background and what was obviously the chaos of Bourbon Street, Moe hung up and was on his way back to us.

The few short minutes felt endless, and I began to panic. I had sent out the rest of the kids to begin looking for my "lost boy." I was geared up to bellow out my best "concerned mommy" voice. But suddenly, through the crowded streets and swarms of drunken partygoers, there appeared half a set of twins and one Monster dog. And once I saw the grin of the proudest boy in the French Quarter, my anger turned. "You did good, son, you did real good."

"Monster helped me, Mom! He helped me find a friend with a phone and then he started walking down the right

street. I think he 'sniffed' his way back." And there it was—
the reason Timmy and Lassie even existed to begin with.
And with my family finally rounded up, we headed out into
the Crescent City to begin looking for our new home.

* * *

In December 2011 we were in a state of chaos while plan-
ning our big move across the country, but we were still trying
to keep the festive spirit alive. It was the last week of the
month, and, true to our crazy family ways, we were all doing
last-minute shopping in between packing our belongings,
both human and dog. The house was covered in unused
wrapping paper, boxes, and bows—and also dog crates, bags
of leashes, and stacks of plastic bins filled with dog toys and
chewies.

And like most kids, mine had parties to attend, friends
to whom to deliver presents, and last-minute wrapping to
do. But unlike most kids . . . mine still held on to responsibilities within the rescue. They got up every morning at the
crack of dawn to feed the 150 dogs, clean kennels, and do
various ranch duties that mandated our survival as a nonprofit dog rescue. And true to tradition, they also knew the
Dog Rescue Christmas Rule: No one was allowed to open
presents until dog duties were completed. It was something
my stepmom had taught me growing up, and I passed that
tradition on to my kids as well.

"It's a privilege and honor to have these animals be a
part of our life. They did not ask for us to take them in.

We chose to bring them into our homes. They cannot get up and open the refrigerator themselves and get something to eat. They have to rely on us to feed, water, and clean up after them. You eat after they eat." Her words still resonated with me year after year.

Christmas morning came and went. Dogs were fed extra goodies and were resting with fat tummies by the time the kids opened their presents. They all gave me quick "Merry Christmas" kisses on my cheek as each one raced out the door to hang out with friends and show off the gifts Santa had delivered to them. I, on the other hand, was left to clean up the mess.

About an hour into my motherly duties, the phone rang. I didn't recognize the local number on the caller ID, but I picked up. The voice on the other end was a young girl's, and I could barely understand her through her hysteria. I think she tried to tell me her name but I didn't catch it. All I heard throughout her constantly repeating her words was: "Moe is in the hospital. We crashed. It was bad." Not to use that cliché, but it was the call that no parent wants to get.

Through the sobbing and shaking fingers, I tried to dial the cell numbers for Tania, Mariah, and Kanani. And the more I dialed the wrong number, my sobbing turned to outright screaming of frustration and anger at myself. I was finally able to hold it together long enough to get hold of all the kids and tell them to meet me at the emergency room, which was a forty-minute drive away.

Despite our location in the high desert, the December morning was extremely cold, and a haze hung over the valley. Most people assume that desert living means hot, scorching days filled with rattlesnakes and tumbleweeds, but when the Mojave is having some downtime, it gets so cold that it hurts your skin. The 363,000 acres of vast, sandy land is unforgiving, and I hated living there. The late afternoon winds came every day like a satanic ritual that took its own sacrificial lambs. But this morning, on December 25, it was still. And with that silvery-colored fog lingering over the "orchards" of Joshua trees, it almost looked like a scene out of a magical *Lord of the Rings* knockoff.

But the reality of it all hit when my kids and I got to the hospital at the exact same time. And there was Moe, in shock, rocking back and forth on a gurney in the hallway with a blanket wrapped around him. My son, the "baby" of the family, was still alive.

As the family rushed to his side, hugging and crying, I could hear his muffled voice cut off by the embracing arms. He was saying something over and over. After the initial shock for everyone had toned down long enough, I held his face in my hands and asked, "What . . . what is it?"

Moe's eyes were filled with tears, and then so were all of ours. "Monster, we need to find Monster. Monster, we need to go and get him." We didn't understand at first, but then the girl who had been a passenger in the car told us what they had endured.

Moe and the girl had left her house after spending some time with her family for Christmas. They were heading back to our house, with Monster in the backseat of Moe's little

Ford Focus. Monster loved riding around with Moe and was very comfortable being a passenger as the duo zipped around the Antelope and Santa Clarita Valleys.

The three of them had come to a four-way-stop intersection that was smack dab in the middle of the empty desert. The fog was very thick and visibility was only about ten feet, but they were familiar with the area, so stopping before they could even see the stop sign was almost automatic. After a brief halt, they proceeded through the intersection—and that's when this Christmas morning turned into the holiday from hell.

As an SUV blasted through the stop sign, it broadsided Moe's car, coming within inches of crushing his legs. The couple in the other vehicle never saw the stop sign until it suddenly appeared through the dense (and ironically monster movie-like) fog. They hit Moe's car full force, knocking it out of the intersection and up an embankment that catapulted the small car up and into the desert sand. The other couple wasn't injured, nor was Moe's passenger friend. My son took the hit.

When the first responders showed up, Moe was still in shock and didn't realize what had happened. And although Moe's friend had gotten out of the car safely, Monster was gone. No one could tell us whether he was thrown from the car out the open window or bolted out the car door as the first responders tried to help Moe. Monster was always protective of Moe, so the idea of his leaving on his own didn't make sense.

At Moe's insistence, we bundled him up and left the hospital, despite some pretty harsh words from the emergency room nurse telling us not to leave. But our family is like that. We love our dogs so much that our life is nothing without them. Despite some bumps and bruises, Moe wanted to go and look for his dog.

With the fog slowly burning off as the morning turned to afternoon, we arrived back at the scene of the accident. Moe couldn't remember exactly what had happened, but working together we were able to somewhat reconstruct the accident based on skid marks, broken glass, and . . . paw prints!

There in the desert sand were Monster's prints. They were a slight distance from where the Focus had finally stopped, showing us that Monster had probably been flung through the open window. In sheer terror as the cars collided, he must have panicked and run off into the desert.

Without any prompting from me, the twins immediately went into tracking mode. They took off into the vast desert on foot while the rest of us stayed on the two-lane highway, driving up and down and calling out for Monster. The landscape consisted of run-down, single-wide mobile homes, many of them probably half-ass meth labs. There were several auto graveyards of cars discarded by would-be car thieves, some of them burned beyond recognition. I remember thinking how many people had told me over the years how beautiful the desert was, but at this moment it was holding my son's dog hostage. I tried not to think about how unforgiving the desert was, but I was grateful that at least it was December and not the middle of August. For this reason alone, I remained positive that Monster would be able to survive for an extended period of time in the infamous Mojave Desert.

My daughters and I were now driving in circles and having no luck. Because it was Christmas morning, there was no sign of life out on the highway, which left us feeling desperate as the hours passed one by one. We had been looking all day for the speckled dog, and I was dreading having to point out to Moe and his brother that it was getting dark.

As we drove back to the scene of the accident, I saw the twins sitting on the side of the road. They were side by side and appeared to be talking to each other. It wasn't until we got closer that I saw what I thought was a desert mirage; it was Monster, sitting between them! As our truck got closer, both boys jumped up and down, yelling that they'd found Monster. As if I wouldn't notice. It was cute.

We did a group hug, with Monster sandwiched between us all. It was only then that I noticed that he and Moe had the exact same injuries: a few cuts on the nose. As we all giggled about it, the twins told us how they'd followed Monster's paw prints into the middle of the desert until they found him under a bush. With the determination and tenacity of a Pit Bull, they wouldn't give up until Moe and his sidekick were reunited.

In the midst of the reunion, no one cared about opening up Christmas gifts or eating a holiday dinner. It had actually started to feel like a normal day for our nonholiday-celebrating family. At that moment our family decided that gift giving was merely a socially acceptable and traditional thing to do. It just didn't mean anything to us anymore, especially with Moe looking at us with this ear-to-ear grin, arms wrapped around his dog's neck, after narrowly escaping death. It was the only gift we needed.

We made several trips across the country to get settled. Moe and Monster were the last holdouts, and finally the whole family was together again. But it came with a struggle. Moe didn't like change or any upset in the balance of his day. He was not comfortable with moving, and we had to all but beg him to join us. After about a month of talking him through it, he finally packed up his best friend, and he and Monster started their trek east on the I-10. Once out of the deserts of California and settled into the celebratory atmosphere of New Orleans, he realized that this was going to be fun.

The TV show flourished in the Big Easy, and the story lines were taking us all over the country now. It was heartwarming to see my kids finally able to experience the world in a more positive light. Watching my kids learn about different parts of the country and the different ways people lived was so satisfying, and I was relieved to see the learning process continue with each adventure.

In April 2013 we were asked to attend an event called the Dog Lover's Day in New York with the Syracuse Chiefs. Between not being into sports and my insane fear of flying, I opted to sit this one out and allow my kids and Earl (the first parolee we hired after landing in New Orleans) to attend the fund-raiser baseball game. It turned out to be a life-changing event for all of us.

After the event the kids reported back to me that they were overwhelmed by the amount of people and support for the cause. They said it was great fun, but they were caught off guard by the massive number of people that showed up.

One would think, having had our own TV show for a number of years, that we would be able to handle public appearances and meeting the fans with ease. But truth be told, my kids and I remain the exact same people we were prior to *Pit Bulls & Parolees*. With me being extremely guarded and somewhat distant when it comes to meeting new people, right behind me are Moe and Tania. They call themselves socially awkward and still to this day do not understand "what the big deal is"; on the other hand, Kanani and Mariah are natural-born social butterflies.

Despite the mob of fans screaming and crying for photo ops, autographs, and requests to "take a picture with my dog," there was one young lady who stood out. She worked with a local Pit Bull rescue group up in Syracuse that was one of the participants at the event, and it was through the mutual connection of rescuing Pit Bulls that Moe, my very reserved and shy son, made a new friend in a girl named Lizzy.

She too owned a Pit Bull, and she volunteered with a group called Cuse Pit Crew. And although I was not at the event, I heard a lot about her through the kids when they returned home. Of course, being the protective sisters, Mariah and Tania were leery and skeptical. With the exposure of the

TV show, it was becoming all too common that people were starting to insert themselves into our lives. So when someone new was trying to infiltrate our pack, the hackles go up. And let's face it, women can be vicious. I mean, why do you think they call us "bitches"?

Within a short time, Moe asked me if Lizzy could fly to New Orleans to stay with us for a few days. Their long-distance phone relationship was ready to move on to the next level, and they wanted to spend some real time together. But more important, they wanted to see how she was around the Monster.

My first thought was: *This poor girl is about to get the shock of her life.* Between our dysfunctional family and our chaotic lifestyle, I was starting to think this could be the beginning of the end for Moe's new flame. I was also concerned because Monster had become increasingly protective of Moe. If Monster had his way, Moe might be sending this East Coast girl packing before she unpacked.

So when I spotted my son and this dog-loving young lady walking down the street, hand in hand, each with a dog—Monster with Lizzy and Jack (Moe's other bad-boy dog) being escorted by both—I knew their first date was a success. With wagging tails and fluttering hearts, love was being unleashed.

For months they traveled back and forth to see each other, and then there were those phone calls that lasted for hours, followed by the constant data overage notices from Verizon that told me Moe was spending more time keeping in communication with his new love. And during this time,

I'm not gonna lie: I was surprised. Moe had never really been the romantic type, let alone the type to stay with one girl for any length of time. So I figured if he could stay with this one for a few months, things would be looking up.

Even though he's my son, I'm the first to admit that Moe can be a little "complicated." He's very straightforward and has no filter. But on the other hand, he loves being a little kid, playing video games, riding his motorcyle, and collecting his Iron Man memorabilia. And for the longest time, it had just been Moe and Monster, with now Jack in tow. He didn't have to answer to anyone but a couple of grouchy old dogs who understood him as if he was their own four-legged brother. They were three "dawgs," just kickin' it and living the single life.

But suddenly there was this female in their life. She had to deal with not only the many faces of Moe but a couple of dogs who really didn't like too many people and had no problem letting everyone know it. But as the months passed, it appeared that Lizzy had managed to make her way into Moe's fraternity. The beauty had tamed the beasts.

On December 6, 2014, I was at the main warehouse helping with the last let-out of walking dogs when Moe called me. Right away I could hear that his voice was almost in a panic.

"Mom . . . Mom . . . I'm gonna do it . . ." Moe had a way of saying only half a sentence or thought. Normally I could figure out the rest on my own, but tonight was different. I was preoccupied with barking dogs, but although I was in a hurry to get the last shift done, I stopped dead in my tracks.

"What's wrong? What? Tell me!" I admit my voice became alarmed. Then it just came out.

"Mom . . . I'm gonna ask Lizzy to marry me. I got the ring and everything!" And there it was. A miracle. Not only had Moe managed to stay with Lizzy for over a year, but he now wanted to settle down. Adulthood never looked so adorable.

And then I went into a state of chaos. I tried to convince him to arrange a romantic setting, any sort of planned proposal, but he was already in overdrive. When Moe gets his mind set on something, he won't budge. This was happening right then and there, and there was no pulling in that retractable leash. All I could do at this point was hang on.

Within seconds I was making calls to the rest of the kids, employees, and even our camera crew. We were scrambling like a Jack Russell Terrier on Red Bull, trying to put something together so as to at least give Lizzy something special for her moment. I mean, we were all in shock. Of all my kids, Moe was the last I would have imagined getting married. Actually, I take that back: I *never* saw my wild child slowing down, period.

Yet there in front of an old run-down warehouse that had survived Hurricane Katrina, in the infamous Ninth Ward area of New Orleans, the "baby" of our family got down on one knee and asked Lizzy to marry him. And through tears and cheers, as she cried out "Yes!" we all quickly chimed in with our overwhelming show of approval. And just like that, my son was "all growed up."

Preparing a wedding can be very stressful for most couples, but for Moe and Lizzy it was more like being party planners.

They enjoyed it and had fun making all the arrangements. And of course, Monster and Jack were to be a major part of the big day. Between deciding what the two dogs would wear, to making sure they were paired up with the right partners to walk down the aisle, Mr.-and-Mrs.-to-Be never had one disagreement leading up to the wedding.

There was talk of Monster and Jack being the best men, but Moe's twin brother, Kanani, became a much more traditional choice. But through all the cake tasting, deciding on venue, dinner choices, and so on, there was no doubt that Moe's best friends were the ones with four legs. But no one could have planned for the next event.

On August 29, with only a few weeks to go before the wedding, Jack passed away from cancer. Jack's reputation as a tough street dog had hidden from us the deadly disease that was manifesting within. That's the thing about these Pit Bulls—they refuse to show weakness or pain. Their big alligator-size smiles stretch from ear to ear, and those whip-like tails are likely to leave welts on the back of your legs. These dogs just have a way of hiding their weaknesses—even to the point of death.

Lizzy actually took it the hardest. She had really bonded with Jack, and his strength in character, body, and soul had filled a void in her life from some of her own family issues. It had become a cute little foursome of "Lizzy and Jack and Moe and Monster," and now a member of their pack had left this Earth.

Lizzy was so devastated that she actually contemplated postponing the wedding. The thought of Jack not joining

Moe, Monster, and her on such a special day made her inconsolable. But we gathered around her as a family and helped her heart heal just a little, enough to stay on track. But now it would just be Moe, Monster, and her. He would be the "lone wolf" to walk down the aisle.

On September 19, 2014, the Madewood Plantation became a living fairy tale. Lizzy and Moe had decided on a *Gone With the Wind* theme. The nineteenth-century southern plantation took your breath away, as its acres of lush green grass unfolded before the eyes. The almost prehistoric-looking oak trees, with their ancient arm-like branches draped with Spanish moss, made the perfect vintage backdrop as Moe walked out with his Rhett Butler–style tuxedo. Even Monster was done up with a themed collar and corsage to match Lizzy and Moe's wedding colors. But there were some sudden unexpected adjustments that had to be made.

Monster's arthritis had gotten worse. He had sadly been bothered by bad knees, but throughout the years we had always managed the pain with meds and holistic remedies. But as time began to creep up on him, his running became trotting, which then became walking. When the walking became waddling, we knew what we had to do.

The twins built Monster his own wagon and taught him to ride in it comfortably. So, on the day of the big reveal and right there on Bayou Lafourche, the dapper dog almost became the highlight of the wedding, outshining the bride herself. And she didn't mind—not one bit.

Pulled by Kanani and Tania, Moe's four-legged main

man put on his best game face and lay in his cart, as proud as could be. Posed alongside the other groomsmen, Monster panted with excitement and anticipation as he, the other wedding guests, and family gazed down the emerald-green grass walkway, waiting for "Scarlett O'Hara" to appear.

As the enormous wooden doors to the plantation home opened, everyone went back in time. Lizzy looked like a forties Hollywood movie star, with her satin, body-hugging gown glistening in the late afternoon sun. And as the music began and she made her way up the oak-framed walkway, there were two men gazing at her adoringly—one with two legs, the other with four. After years of looking out for his boy Moe, Monster was now ready to accept the new woman in his life as well.

As Lizzy neared her husband-to-be, the fatherly figure of a dog slowly wagged his tail in a gesture of approval. And although Jack was with them in spirit—an enlarged framed photo had been placed in the after-party area—the newlyweds were just grateful that their "Monster Man" had hung in there long enough to be a part of the next chapter in Moe's life.

His old body was worn out, and with each movement you could almost hear the creaking in his wise and tired joints. You could see in his face that the pain was becoming unbearable. With Lizzy and Moe having to carry him down the stairs and outside, he was appreciative, but his pride was getting the best of him. Monster knew it was time. It had been ten days since the wedding, and Moe now had a wife to help take care of him. "Mi-Ya" knew his job was over.

He had protected the baby of our family for all these years and had gotten him through the darkest time of his life. Moe had gone from a lost and confused boy to a man who had found his way. He was now Mr. Chock, and much stronger emotionally. And in that moment, Moe knew that it was time for him to step up and accept the responsibility of what "I do" meant.

Right then, the boy and his dog became the boy and his girl. Together, hand in hand, they made not only their first decision as a married couple but maybe one of the hardest ones they would ever have to make. Mr. and Mrs. Chock stood by the dog who had been through so much yet remained so strong. And as Monster took his last breath, both Lizzy and Moe leaned in and whispered the words, "Till death do us part."

Back at home, the closet door slowly eased open, and the boy in the dark was no longer alone. His girl lay next to him, and their arms embraced each other with a knowing smile. Gone was the fear of the bogeyman and things that go bump in the night. The blankets were no longer pulled up and over their heads but, rather, wrapped around themselves in that typical newlywed snuggle.

As Moe turned off the bedroom light, he realized that the demons from his past were gone. And as he slipped back into bed, next to his new wife, his eyes still adjusting from the light to the dark, he swore he saw something . . . someone peering from within the pile of his work boots and disorganized hanging clothing. He smiled. He knew. Sweet dreams, Monster Man.

7

LUCKY

The One-Woman Dog

My stepmother, Margaret, taught me to be strong and to not rely on anyone. Thanks in part to her, I learned how to do all kinds of "men's jobs" at a young age: change tires, do minor mechanical repairs, fix broken plumbing, and repair fences. It felt great to be self-sufficient, but I didn't spend much time wondering why Margaret had considered all of that so important.

When I think back on it, I realize that she was motivated—at least in part—by the pain my father had caused her. She didn't want me to ever *need* a man, though *wanting* one was perfectly okay.

My own first heartbreak came when my high school sweetheart Jim abruptly dumped me after graduation and ran off with a cheerleader. It probably sounds like a cliché now, and no big deal, but I promise you—it *was* a big deal. When I close my eyes, I can still summon up the feelings of that early morning, sitting in my 1965 Chevy truck sobbing uncontrollably over the boy I thought was my "one and only."

That breakup made me a little gun shy, I guess. I didn't have many long-term relationships during my younger years and, in fact, I really haven't had that many to this day. It's partly due to lack of trust but also, I have to admit, I wasn't always the best of girlfriends. Not that I was a cheater or a party girl, but my "abnormal" relationships with my horses and dogs tended to be a turn-off. Who wants to compete with all that—and who wants to date a tomboy, anyway? Yeah, that's what they called it in those days, and the label didn't really bother me, especially if being a *tomboy* meant I could wear my favorite baggy boys' T-shirts and "roper" cowboy boots day in and day out.

As a young adult, I finally discovered the male species and, of course, there were several more heartbreaks over the ones that at the time I felt were "true love." I actually went through phases when it came to men. I had my truck driver phase and my cop phase. Then there was my musician phase . . . and, thinking back on it, I left a trail of some broken hearts myself. But my rebound relationships always fell to my companion animals, specifically my dogs.

Cougar got me through some pretty tough times. She was a little Catahoula/Pit mix bursting with bravado and a little too big for her britches. I guess you could say she was my first "mature" dog relationship, and she paved the way for many more to come. I still carry around a mountain of guilt about never getting to the bottom of what happened to her when I went off to the military. My father offered only the oldest of explanations—some variation on "We sent her away to a farm in the country"—and I was eager to believe it. Today, after years in the animal rescue business, I know exactly what that means . . . and I still kind of hate him for it. Eventually he would admit that Cougar was sent to an animal shelter.

As the years came and went, there were periods in my life when I was dogless. Not for any length of time, though, since I just couldn't survive emotionally for long without a canine at my side.

But as a young, single mom, my life wasn't always stable. Actually, it wasn't stable at all. So whenever I found a stray or took in someone's unwanted dog, I would eventually end up having to find a new place for the homeless pooch. It was hard enough just keeping things together for me and my oldest daughter, seeing as how we were destitute and living out of my 1966 Volkswagen Bug.

We had our ups and downs during those years, but eventually I gained some stability and, with that, my passion for dogs became full time.

I've heard you're supposed to "follow your passion" to find a career path, and, though it sounds a bit frivolous as a description of my life, there's no denying that my passion for dogs has defined my life's work. For better or worse, it has carried me from that VW to where I am today: an ardent animal rescuer and the world's most reluctant reality TV star.

Villalobos Rescue Center soon became my entire world. And during that time, my human relationship with my second child's "baby daddy" lasted for twelve years. He was supportive of my rescue efforts and we shared a love of dogs—but, in true Tia Torres fashion, my life literally revolved around my dogs and, of course, my kids. My common-law relationship ended on Christmas Eve and I found myself relying on the four-legged man in my life, Duke.

Duke, my red-hot prince, passed away in 2007 and left a huge hole in my heart. He would be very hard to replace. In my head, I knew it was foolish to try to find another Duke, but let's face it: My life was a Pit Bull smorgasbord and I was hungry. Every time an oversize, red, crop-eared dog came through our gates, I was "in love" . . . until I wasn't. I finally realized that I wasn't ready to love a Duke clone and vowed to practice what I had always preached to my kids: Lay back. Love comes when you least expect it, and from the most unlikely sources.

That said, Tia has a type—don't we all? Call me superficial, but I have always been attracted to tough-looking dogs. I think what I find so appealing about them is that no matter how scary they appear on the outside, they tend to be big puddles of mush on the inside.

In any case, after Duke left this Earth, and after interviewing hundreds of "applicants" for his replacement, I gave up on finding a new four-legged mate, at least for a while. That's about the time when a little whirlwind of a life-changer called *Pit Bulls & Parolees* entered my life. Without realizing the magnitude of the change in our lives this would cause, my kids and I agreed—for the sake of the dogs—to put our work and our lives on display. And, a few years later, we found ourselves moving across the country

when a gaping need for us arose in the wake of that stormy
bitch named Katrina.

★ ★ ★

Moving from L.A. to New Orleans was a complex operation, involving multiple exhausting cross-country trips, but
I managed to get the last group of dogs there on January 1,
2012. We hit the ground running and were rescuing dogs
before the California pack had even settled in. We had no
choice! The calls were pouring in like rain through an open
roof, and I started with the one that seemed most pressing.
Apparently, there was a "very sick dog" loose in the devastated Lower Ninth Ward.

The twins, Tania, and I jumped into our vehicles and the
camera crew followed close behind as we made the drive
over the bridge that has now become a common sight on
our show. As we turned down a street that ran along the
industrial canal, I spotted a dog out of the corner of my eye
and yelled for everyone to stop. Was this tan-and-white Pit
mix with floppy ears the dog for whom we were looking?
That dog had been described as "barely alive," but this guy
didn't appear sick or even scared. In fact, he was happily
peeing on a bush when I got my first glimpse of him.

"This can't be the dog," Tania said, but he needed rescuing all the same, so I jumped out of the van with some treats
and called out to him. No dice. He just kept roaming from
bush to bush, lifting his leg, even after he had nothing left
with which to water them. I was torn. I didn't want to leave

this cutie behind, but Tania kept reminding me that the sick pup wasn't going to rescue himself.

I want to be clear about something here: We never do anything just for the sake of the cameras. That was my firmest condition when I agreed to do the show—that we would go about our business as we saw fit, and the camera people could film what they wanted. No staged shots and no retakes. *But* . . . there are times when it does cross my mind that something might work out great for the show—and rescuing two dogs in the Lower Ninth for the first episode of our first New Orleans season was one of those.

Seeing that my dog treats were of no interest to the tall Pit mix, Tania tossed some leftover Kentucky Fried Chicken out the window. He sauntered over to it, sniffed, peed again, and then—more like a stag than a dog—leapt over a garden fence and then another and another, leaving us in the dust. Clearly, he had no interest in being "saved." Shrugging my shoulders, I turned back to the van, only to see Tania and the crew smirking up a storm.

"What an asshole dog," my daughter snorted. "You know how to pick 'em, Mom."

What could I say except "Let's get going"? We drove on and rescued the dog we'd been called about—named Gator—and got some great footage for the show. But I didn't forget about the pup who got away. The Lower Ninth Ward was not a good place for a stray dog; you had to be one tough pooch to survive there.

About a week or so later, coming back from my weekly trip to Walmart, I rolled up to the stoplight where the Martin Luther King Library meets Judge Perez Drive. As the light turned green, something on the curb caught my eye—a flash in the bushes that I quickly figured out was a dog. Ignoring the green signal, I rolled down the passenger window and tried to get the pup's attention.

Using the high-pitched doggie-calling voice with which all pet owners are familiar, I called out, "Hey, buddy! Hey, you!" An impatient motorist behind me leaned on his horn and I motioned for him to go around me so I could stay focused on the peeing pooch, who was now looking up at me. Wait . . . a peeing pooch?

Yup. There he was. That same unappreciative, floppy-eared Pit mix who hadn't given a rat's ass about being a reality TV star. Partly out of frustration and partly because I meant it, I yelled out, "I just want you to love me!" And with that, he took off down a side street. My ghost dog with a 'tude just trotted away from the crazy lady pouring her heart out through her car window and disappeared.

I came back to the rescue ("the warehouse," as we call it) and told the story with arms flailing in an attempt to describe this dog's "Flying Nun" ears. The film crew was ready to jump in their ever-so-cool-looking minivans, but with a defeated drop of my head, I told them that this dog did not want to be caught and didn't give a crap about anything I had to offer. As much as I hated to walk away from a dog in need, this canine had his own agenda and I wasn't on the list.

A couple of weeks passed and phone calls and emails continued to pour in about dogs in need. News of our presence in Louisiana had spread fast, and we were honestly overwhelmed. In three seasons of filming in California, we'd never experienced the volume of distress calls with which we were currently faced, and I had to figure out how we could meet the need.

For eighteen years, I'd survived the cruel desert terrain of the Antelope Valley, enduring bone-dry 110-degree days surrounded by dirt, rattlesnakes, scorpions, and 75 mph Santa Ana winds. Southeast Louisiana was a whole different story, starting with the heat, which was as wet as California's had been dry. I was told I would get used to it, and eventually I did stop sweating like a you-know-what in church, but it made chasing a dog extremely exhausting.

The snakes were different too. Unlike rattlers, who send you a warning when you're getting too close, the serpents of the South are stealthy creatures. Believe it or not, I had gotten so used to rattlesnakes that I didn't think twice about scooping them up with a broom or shovel handle and tossing them out of my dog kennels. Louisiana's thick black cottonmouths were another story. I'd need a little more time before I'd snuggle up to one of them.

Then, of course, there were the alligators, which sometimes made their way into the canals of the Crescent City. But who would've imagined that our greatest enemy would also be the smallest: mosquitoes, which carried heartworm to 99 percent of the dogs we rescued.

Talk swirled around town about our chances of surviving all these challenges. Thankfully, my family was used to constant change and never-ending obstacles. Come to think about it, in some twisted way, we thrived on them.

★ ★ ★

In mid-March a call came in about a dog in a very peculiar situation, and we jumped at the chance to get involved. The story was a long one and sounded a little bit like a Disney movie waiting for its happily-ever-after ending.

The caller was a mom, who quickly explained that each day when the school bus dropped off her young daughter, Dee, the neighborhood stray dog—Beaux—was waiting for her. Beaux didn't belong to anyone that she knew; he just hung around the local yards, and everyone took turns feeding him. He was a small, brownish dog with one torn ear, she said, and each day, like clockwork, he would wait for little Dee and escort her right to her doorstep. That is, until the day in question.

On this day, as the storybook pair walked down the street, Beaux began to act strangely, barking incessantly and spinning in circles. He kept running back and forth to the field that paralleled the bayou. At first, Dee didn't think much about it; then Beaux took off across the field, straight toward the edge of the embankment that led down to the water's edge. Barely able to see his little head over the tall grass, Dee could still hear his high-pitched barks and grew afraid of what would happen. She knew this wasn't normal behavior for her after-school escort.

Dee dropped her book bag and made her way down toward the bayou, following Beaux's increasingly frantic barks. When she could finally peer down into the muddy water, all she saw was the usual array of ducks, egrets, and cranes. Then, looking a little farther out, she saw something that didn't belong there. She knew in an instant that this was the cause of the uproar: a drowning dog.

The somewhat large dog was tangled in an enormous sea of swampwater lilies. His front end was hanging on for life and he was panting heavily, struggling to stay afloat in the heat and humidity. Dee remembered a time when, as a younger girl, she was fishing out of this bayou and accidentally caught a gator—so the reality and the urgency of the situation quadrupled in her mind.

As her mom told it, Dee and Beaux came bursting into the house and physically pulled her down to the water's edge to show her the dog, who was showing signs of tiring out and giving up. They immediately called the sheriff and animal control, both of whom arrived quickly and scaled down the embankment.

No dice. The dog was too far out into the water to be reached without a boat—but they knew there was no time to get one. With nothing to lose, the animal control officer made a Hail Mary toss of a rope and managed to loop it around the dog's neck, then slowly began to pull him to safety—a difficult proposition to say the least, because of the tangle of plants and muck. Luckily, the dog was too tired to do anything but offer a few half-hearted kicks with his exhausted back legs.

When the waterlogged pooch was finally safe on the bank, he received quite the welcome. Beaux, of course, was ecstatic, as was Dee. Other neighbors had gathered to hear the story and marvel at everybody's quick thinking and hard work to save the lucky animal. Beaux and Dee were heroes.

But now another search was in order: They needed to find out if this dog had an owner. Reluctantly, Dee and Beaux watched as the exhausted dog was loaded up into the animal control truck and taken to a shelter.

And that's when Dee's mom called me. I agreed to take the dog if no owner came forward—and a week later, I got the call. As I made my way over the bridge on St. Paddy's Day, camera crew in tow, I wondered for the thousandth time how this poor dog had ended up in the bayou. Did he fall down the embankment? Did someone throw him off the small bridge that connected Dee's neighborhood to the next one? I knew I'd never find out, but one thing was clear: This pup was a survivor. There was no way to know how long he'd clung to life, but he clearly had a will to live and was one tough boy. I couldn't wait to meet him and express my admiration directly.

I walked into the shelter with my entourage, where we spoke with the shelter staff and prepared to meet the miracle pup. This took longer than I wanted it to, because we had to get the microphones and cameras in position, and on and on—a process I had grudgingly gotten used to by this point. After what seemed like a very long fifteen minutes of prep, we were finally escorted through the doors to the kennel area.

Passing six or seven other barking dogs in cages, the kennel attendant stopped, pointed, and proudly proclaimed: "Well, here he is!"

My heart pumping, a smile nearly splitting my face in two, I knelt down to give this dog the hero's welcome he deserved and then . . . the needle scratched across the record album. I could not believe who sat before me.

"You!? Are you effing kidding me???"

I stood up and stepped back, and my chuckle turned into a belly laugh. Sitting there, expressing no emotion whatsoever, was my pissing dog with a 'tude; the heartbreaker of the Lower Ninth Ward. And even now, the dog whom I'd chased and courted and begged to love me was giving me the cold shoulder. There was no "happy dance," no "take me home." He just sat there with a look that said, "Bitch . . . what the hell do you want *now?*"

The kennel worker delivered his best sales pitch, insisting that the dog couldn't be sweeter, and *blah, blah, blah.* When I could get a word in, I told him he didn't have to worry—we were taking the dog no matter what. For one thing, he was about to be the star of a season-four episode of our show titled "Ghost Dog."

I was content to take my cool customer straight back to the warehouse, but the producers wanted to film a "happy puppy frolicking in the grass" scene and asked if there was somewhere they could do that "really quick." Something told me that frolicking wasn't this guy's specialty.

It took us another fifteen minutes to coordinate everything out in the exercise yard, and I spent the entire time watching a kennel worker lead this dog out to "freedom." It was a really nice yard, with agility obstacles placed around. And just like that, the "Hound of the Baskervilles" was on the loose. With that same attitude he had exhibited on the streets of the Lower Ninth, he strutted about the fence line, stiff tail straight up, stopping to "mark" everything in sight, including a couple of crew members.

This dog carried himself with arrogance, paying absolutely no attention to the group of people standing there watching and filming his every move. He had his own agenda and could not care less about anything but himself. After making sure his territory was properly labeled, he sauntered over to the hanging tire obstacle and nudged it with his nose. That got him a couple of giggles.

Then he grabbed hold of the tire and began to tug . . . really hard. This brought forth an explosion of laughter from the crowd of spectators—with the exception of me. This dog's body language was screaming everything but "cute and funny." Although I didn't get the sense that he was going to act aggressively toward any of us, I knew his overly confident demeanor was something of which to be wary. I told the crew to give him some space and not engage with him in any manner.

It was at that moment that our newest charge yanked the entire wooden frame and tire out of the ground and began swinging it around until the two-by-fours were reduced to splinters. He then took off at full speed, dragging the contraption with him and even slamming it up against his own head.

Finally, the entire frame fell to pieces and my new pooch dropped what was left of the tire and became more interested in a remaining foot-long chunk of two-by-four. Picking it up like a twig that had been tossed to him in a game of fetch, he began racing around the yard, then headed—deliberately, I thought—straight into the crowd, striking people in the knees, shins, and various other body parts. Camera people, producers, and staff scattered as if Godzilla had just infiltrated the city.

And then he stopped. His job, whatever he'd perceived it to be, was done. He dropped his weapon and lay down, panting with a combination of exhaustion and contentment. This was no ordinary dog. I'd have my work cut out for me with this guy.

Like most dogs that we take in, we put him into a kennel at our main warehouse. I had asked Dee what she wanted to name him, which seemed fair since she was his savior, and she didn't hesitate. "Let's call him Lucky," she told me, and I had to admit it was perfect—even more so because I got him on St. Patrick's Day.

By the next day, I was getting calls from my staff that Lucky was wreaking havoc in the warehouse. The kennel workers couldn't even get close enough to his enclosure to feed him. The fact that he was showing some serious boundary and guarding behavior was not completely shocking. I had always felt that there was something about this dog that would challenge my experience. Now I would get to meet the real dog.

For the next couple of weeks, I made Lucky my personal mission. I fed him, cleaned his kennel myself, all the while wondering, *What have I gotten myself into?* There was no connection between us and he exhibited zero emotion. Whatever familiarity I was building with him was unlikely to end in hugs and kisses, and my plan was just to get through each day with him.

As the weeks passed, Lucky didn't mellow one bit; in fact, he grew increasingly aggressive toward everyone but me, so, against my better judgment, I took him to my house, where I already had quite the pack of antisocial misfits. I wasn't looking forward to taking on another—but I did have that weakness for bad boys.

In fact, Lucky was pretty polite and did everything I asked of him. I discovered that he was simply one of those dogs who is on guard all the time, hyperaware of his surroundings and on high alert for even the tiniest threat. He oozed self-confidence, which is a pretty attractive quality, at least to me. Quite simply, Lucky made me feel safe—and for that, I was a little bit in awe of him.

Lucky pretty much ignored me. He ate when I fed him, went outside when I asked him to, lay down when he was done with his routine. It had been five months, and not only was Lucky still indifferent toward me, he was proving to be more than aggressive with other people, continually trying to attack them. Any thoughts I'd had of finding him a home other than mine had vanished—though I certainly didn't need yet another aggressive dog with whom to contend.

Living in a single-wide trailer out in the swamp proved as agreeable to Lucky as it was to me and my other two "problem children," Arnold and Marvin. There was nothing out there to rile him up. Our large yard with beautiful green grass was a particular pleasure for me, after all those years in the desert. Life was serene all the way around, but I was haunted by what I feared Lucky was capable of doing. I knew I had to try to work on his behavior, and I knew it wouldn't be easy.

At this point, I did know a bit more about what made Lucky tick. Shortly after his miracle rescue, the local newspaper had run a story about the brave little girl and her dog who'd saved another dog from drowning. After the story appeared, from out of nowhere, Lucky's former owner surfaced. Apparently Lucky had been trained as a guard dog using the "agitation method," which, in my experience and opinion, is not the safest way to train a dog for protection work. This gem of an owner was quite happy to tell me about his training technique. He'd tie Lucky to the door handle of his truck and then flick a towel or rope at him until he got agitated enough to attack. It got to the point where he was "so well trained" that any movement whatsoever would set him off and he'd lunge. Everyone became his enemy.

I had already stopped taking Lucky out in public, and this information confirmed all of my suspicions about him. He had been taught to be a dangerous dog. And now his former owner wanted him back.

There was no way I was going to allow that to happen. After a few expletives and a lot of screaming on my part, the man left with his tail between his legs, and I was stuck with an angry dog who needed deprogramming.

We were deep into our first summer in the South, and, although the humidity made the air thicker than a McDonald's milkshake, I discovered the pleasure of the swamp at that time of the year. The river was alive with critters and lined with beautiful plant life I'd never seen before, and the night sounds were unlike anything I had ever heard. I had been accustomed to the chirps of crickets, but here they were joined by the grunts of bullfrogs and harrumphs of alligators and a lot of sounds I couldn't even identify, all of which soothed me. It was also hurricane season, but that was just an abstraction to me at the time.

It had been five months since Lucky was pulled from the bayou, and he still had to travel everywhere with me because nobody else could get near him. My kids and I were in Texas for a public appearance at a tattoo convention and we were scheduled to be there for three days. Just a few hours after we arrived, we heard about a tropical storm brewing in the Gulf of Mexico.

Although Mother Nature has been my longtime arch nemesis, a hurricane was not something she had ever thrown my way. The locals were cautiously optimistic that Isaac, then a tropical cyclone, would just "go away." I, on the other hand, was worried. This was a first for me, and Tia Torres does not like feeling vulnerable.

By Sunday, the convention was winding down, and one of the best artists there offered to do a tattoo for me. For years, I had wanted to pay homage to my godfather but could never find the right person for the job. Now, here he stood, ink and gun in hand and with the chops to properly immortalize Lon Chaney Jr., the original Wolf Man, with a crowd watching his every stroke. The entire piece took a few hours to complete, and we were in the home stretch when someone at a neighboring booth yelled out the words that no one in the South wants to hear: "Hurricane on the way!"

I looked across the aisle at Tania, who was also getting a tattoo done. I will never forget the look of alarm on her face. Trying to remain calm for her sake, I told the tattoo artist, George, that I needed to go. He quickly cleaned up the excess ink on my arm and left me with a strongly worded warning about keeping my new piece clean and dry. "Don't expose it to anything nasty or it will get seriously infected," he shouted after me, but my mind was on another kind of disaster, so I barely heard him.

Lucky and I would travel separately from my kids. We all had our dogs with us, and no one but me could ride with my new sidekick. We quickly dragged all of our luggage, dog crates, and dogs out of the hotel and set out on the longest six-hour drive of our lives. Lucky rode shotgun, curled up on the seat acting like he was asleep. But, true to his suspicious ways, he kept one eye on me as though he knew something was brewing.

Hour after hour, I switched back and forth between news stations, skipping over my usual country-western favorites. As I listened to the reports of the storm growing stronger and bearing down on our area, I gripped the steering wheel so tightly I thought I might snap it in two. Every so often, Lucky would raise his head and stare right into my soul for a few seconds, then lay his head back down. It seemed as if he was trying to assess a situation that neither of us understood—nor did we realize how much we'd have to rely on each other in the hours to come.

Both vans arrived back in New Orleans to find everyone at the rescue in full 911 mode, the ghost of Katrina hanging in the air like a fog. Someone had pulled out the old school bus we'd used in our move from California, and some of the guys were filling it up with crates. Our hurricane evacuation plan had always been to remove as many dogs as possible to the little town I lived in outside the city, so every van, trailer, and other vehicle was being prepared in case it became necessary to evacuate them.

We all had the horrors of Katrina on our minds and were understandably leery of government officials saying "Don't worry. The levees have been improved and will hold up." Although we hadn't been in New Orleans during Katrina, we'd witnessed the immediate aftermath, when, during the weeks that followed, we took in sixty Pit Bulls. With that on our minds, we opted to move some of the dogs sooner rather than later, to the little town of Maurepas.

Once the designated dogs were loaded up, I instructed everyone to begin moving the rest of them upstairs at the warehouse. The storm was due to hit within a day or two and the city was getting ready to go on lockdown, meaning no one would be able to enter or leave via the interstate. As the staff raced around, I heard a sound coming from my van and realized Lucky was still inside, now raging and throwing himself against the window. I'd jumped out so quickly that I'd left him there, in the middle of the chaos, and it was making him crazy.

By the time I walked over to him, he was in such a state that he didn't even recognize me through the window and continued to bang his head and throw his spittle around like a madman. As Tania raced by, she said, "Mom! I don't know how or why you deal with that guy!"

And she was right. We were in the middle of a crisis, and my dog wanted to kill me. No one else could get near him but me, and even that was becoming a challenge. It was so debilitating. Although it was not the time to be making any harsh decisions, the thought of sending Lucky to "doggie heaven" was beginning to feel more reasonable. For now, I had to just deal with him and handle business.

I quickly yanked the door open and, in a firm, but calm voice said, "Hey! Relax!" And just like that Lucky jumped back over onto the passenger seat and lay back down. I had quelled the smaller of the two storms, but something told me Isaac would not be as compliant.

As Mariah and I, along with a handful of hardcore and dedicated workers, got ready to make the hour-and-a-half drive out of the city, we felt confident about our plan. We'd left behind only as many dogs as would fit upstairs at our main facility. But the fear of something similar to what had transpired during Katrina still sat heavily on our hearts and frayed our nerves. It had been seven years since that Category 3 hurricane had hit, yet there was a sense that it had only been days.

We arrived at my trailer before dark and unloaded my own dogs and a few others from the bus. Mariah and the team then took the other dogs to a few locations we'd prepared for them and set about hunkering down for what was to come. Once everything and everyone was in place, it was just a waiting game, as the weather reports all confirmed that the city was in the direct path of Isaac.

My own thoughts turned to the various natural disasters we'd endured in California. There had been the yearly wildfires during the summer months and the blizzards in the winter—one of which froze everything so solid that we had to break the ice in the dogs' water buckets for days. And every year from October to March we'd dealt with the Santa Ana winds, which got up to 75 mph and more nearly every night, wreaking all kinds of havoc that we then spent our days repairing. Compared to all that, I figured a little Category 1 hurricane would be endurable—if a little wetter.

Lying on the floor of my trailer watching the news, I learned firsthand the meaning of the phrase "the calm before the storm." It was just so damn quiet that I got restless. I paced around, looking out each window in turn as the skies darkened. We had rented a sturdy brick building downriver just in case we had to move dogs out of my trailer and the trailer where Mariah was staying with a few others—as well as the now TV-infamous "Cheech and Chong house" just across the river. The locals had assured us that the area we were in hadn't flooded in over fifty years, so we were cautiously optimistic as we waited for Isaac to bring it on.

The storm hit—or, more accurately, took up residence—for two days, pounding us relentlessly. The dogs seemed to handle it pretty well. Lucky, of course, remained as cool as James Dean in a street fight, even when the trailer shifted and shook so hard it seemed it might tip over. I'm not gonna lie: This human was scared.

On the second day, the power went out and it was time to make a move. With the last bit of charge left on my cell phone, I called Mariah and our two colleagues across the river and made plans to move the dogs to the evac center upriver.

We loaded up the Jeep Wranglers—the only vehicles we felt could do the job—and took off into raging wind and rain so heavy that we could barely see five feet ahead. Surprisingly, the roads were still passable, as long as we steered clear of the few trees that had fallen across our path.

We left two of our workers to stay with the dogs, who were pretty unsettled after being evacuated twice in the span of a few days. But that didn't solve all of my problems. I still had three dogs in my trailer who couldn't be around humans—Arnold, Marvin, and Lucky—so I sucked it up and decided to ride out the storm with them.

The night was a long one. The trailer continually rocked ominously, and the battering rain on the windows sounded like strokes of a baseball bat. I was rethinking every decision I'd made, especially the one that had brought us to Maurepas. With no power left on my cell phone, I couldn't even speak with New Orleans to see how everyone there was doing. I sought reassurance and companionship from the dogs, but Lucky and Arnold seemed determined to sleep right through the festivities. Marvin, on the other hand, was a wreck—worse off than I was—so he wasn't much help. My only illumination came from a little camp light that I was trying to conserve. So typical of Mother Nature's twisted love/hate relationship with me, her worst was yet to come.

Eventually I guess I fell asleep on the floor, and when I woke up, the sun was peeking through the curtains. Isaac had moved on. But before I could register any relief at the thought, I noticed I was wet. Soaked, in fact. I slowly sat up and surveyed the situation. I could see that all down the hall and into the living room, the carpet was drenched. My first thought was that it was a broken pipe, but my trailer was elevated about four feet off the ground. If a pipe had broken, the water would've poured onto the ground, not into the trailer.

When I looked out my window, I screamed out loud, waking the dogs in their crates. The entire town had flooded from a storm surge. The water was up about five feet in some places and all the way to the rooftops in others. People were out in boats, and I could hear the echo of generators buzzing throughout the swamp. I walked out to my porch and found myself standing in about a foot of murky water. My "dog van" (later resurrected as the Beast) was in water just up to the bottom of the doors. Because the Jeeps were "lifted," there was still some wiggle room to go before their interiors were flooded. But the enormity of the devastation was overwhelming. Panic began to set in as my mind raced.

My first thought was to get my dogs to safety—but how do you get three ornery animals who don't get along with one another or anybody else into a Jeep Wrangler? I let each dog out of his crate to wade around in the increasingly flooded trailer. In the short amount of time since waking up, the water had risen a few more inches. Ankle deep, each dog squeaked a little as he minced around, trying to find a dry place (which turned out to be my antique sofa), while I lifted their crates up and onto my kitchen counters. After ten minutes of slipping around after them in the swamp water, I managed to get all three of them secured and dry.

I told them that everything would be okay—though I don't know if any of us believed it—and opened the trailer door to a mini-tsunami of muck that poured into my living room. I looked back and had to give Lucky a grim smile. He was chilling peacefully, as if it were just another day in the life. His head rested gently on his front paws as cool as you please. For the hundredth time since I'd known him, I thought: *If only I could be like you.* This time, I added: *If only you were human, maybe you could help me figure out what to do next.*

With one last wave at the awesome threesome, I waded across to the Jeep, got in, and crossed my fingers. She started up with no problem—go, Jeepy!—and not a moment too soon. The water had started to creep in through the bottom of the door as I headed off to try to find Mariah.

I was in the process of putting the Jeep in reverse and attempting to back out when I heard a beautiful sound: "Mom! *Mommeeeee!* There she was: my youngest, rowing herself, a friend, and her dogs in a big bass boat, grinning ear to ear as she made her way toward me. She was drenched and her hair was sticking to the side of her face from sweat and river water, but she made me smile. "We did it! We got out," she cried, her tiny voice pinging off the huge cypress trees as if she were a little wood nymph. Despite her tiny stature, Mariah is a giant when it comes to determination, and as resourceful a person as you'll ever find.

Her trailer had taken in about three feet of water, so we headed back to mine to get the dogs out and make our way to the evac center upriver. By the time Mariah managed to maneuver the big boat back to my place, the water had risen another two feet. Thanks to her fragile rapport with Arnold and Marvin, we were able to get them settled in the boat fairly easily, despite the fact that they'd been terrified out of whatever wits they had left.

Then it was time to deal with my baddest bad boy. As I waded through the water toward my trailer, I looked down at my arm and realized that my fresh tattoo had been submerged in toxic sewage for quite a while. *That can't be good,* I thought, but my mind was on saving my final four-legged friend, and the possibility of losing the memory of my godfather was just going to have to wait. (Within a couple of days, I'd learn all about the fun of a staph infection.)

And then there was Lucky, who, for whatever reason, trusted only me. Like the tough dog I knew he was, he jumped up in the little wooden homemade bateau boat that someone had made for me and I got into the chest-deep water and pushed the now soaking-wet animal out through the yard and toward the highway where the water was not as deep. And just like that, Hurricane Isaac became a matchmaker. To this day, I don't know who took the photo, but the image of me and Lucky became symbolic of what we went through—together.

For years after that, Lucky and I were inseparable—and I mean we were never apart for a single night. After Isaac, we bounced around from hotel to hotel, rental to rental, and even lived in an RV for a while. He and I and the rest of my incorrigible pack just couldn't seem to put down new roots. But no matter where we were, Lucky was content as long as I was there. I think it is safe to say that we were both suffering from a wicked case of PTSD, and the thing that pulled us both through it was each other.

I began traveling all over the country for the show and brokering adoptions on camera and off. Eventually, my extended family outgrew our passenger van, so we stepped it up and bought an RV—but not just any RV. We got a 1999 model, straight out of Picayune, Mississippi. Nothing but the finest for "Trailer Trash Tia."

What with our hectic filming schedule and the fact that we'd moved back to the swamp (a different one) two hours outside of New Orleans, Lucky and I spent quite a lot of time in the RV. Maybe it seems strange after our hurricane ordeal, but I continued to prefer my "swamp shack" to the city and would only drive into town when I had to. After performing whatever tasks were at hand, I'd hurry back to the "swhetto" (where swamp meets ghetto). Lucky seemed to agree with me and took to lying on the dashboard as we drove, glaring at anyone we passed.

It had been three years since that first St. Patrick's Day, and my boy was becoming a man. He'd settled down somewhat, but he still enjoyed intimidating anybody he perceived as a threat to me.

On January 15, 2016, we were preparing for a big fundraising event at our bar. We had booked the Grammy-winning band Confederate Railroad, and everyone was running around like nuts getting ready for the night's festivities. I, for one, couldn't wait for the band to hit the stage and sing my theme song, "Trashy Women." It was one of those rare opportunities for me to forget my responsibilities and let loose.

Earlier that afternoon, I had loaded up my hounds and Lucky into the RV and driven them into the city. We'd parked in our usual spot, under the overhang in the parking lot at the production office. This was my makeshift campground, complete with a view of Poland Avenue's nonstop traffic. It wasn't exactly a cabin among the redwoods, but it was life in the Upper Ninth Ward.

I spent the afternoon running back and forth between the bar, the rescue, and the dogs hanging out in the RV. They'd come to love it in there, even more than being at home in the swamp. That night, I made sure to walk them a little longer than usual so they'd be tuckered out—which would allow me a little more time at the concert to get tuckered out myself.

I got to the bar in the early evening because we had offered up a small meet and greet for fans and supporters. Although I am extremely appreciative of what we mean to our fans, it is still a little awkward for me to accept the role of "reality star." I managed to make a little speech to the crowd, then posed for pictures as they'd come to expect. My ongoing problem was that I felt like the most reluctant celebrity on the planet. My kids and I had never cared about being famous or even being on TV at all, but we'd agreed to let a crew document our work, viewers had taken a shine to it, and we were stuck with the consequences.

Since, as I said before, our show is 200 percent real, it's hard for me to think of those of us who appear on it as *stars*—but, then, there are lots of things about the entertainment world I don't understand. People look at us differently than they used to, and I have to resign myself to it—but doing the PR that is part of my job description will always feel weird.

The time had finally come for me to meet Confederate Railroad. We were in what we called the VIP room at the back of the bar. There were just a few of us there, and the twins and some of the other employees were posing for pics. I had gotten caught up on a phone call and was sitting on my favorite "white trash" couch. As I looked over my shoulder, the twins motioned for me to come over and pose for photos with the band. I gave them the "in-a-minute look," not realizing what a fateful move I was about to make.

I finished the call and shoved my phone down the front pocket of my overalls.

"C'mon, Mom, get in this!" Moe yelled, and for some reason, instead of simply getting up and walking around the corner of the couch, I decided to take a short cut. I put my left foot up on the back headrest and pushed off to swing my right leg over. It was only about a two-foot drop, but as my right foot hit the ground, my left one got stuck and slipped between the cushions, twisting like a cyclone. The "snap" could be heard throughout the room. As an uncomfortable groan emanated from the crowd, I knew I was in deep trouble.

The twins grabbed me as I fell forward, and I was able to regain my balance. As I put my left foot down, I felt instantly hot and flushed, and Kanani and Moe must've seen me start to go. Each one grabbed an armpit and they held me up. *Damn it, sprained ankle,* I thought, balancing on my right foot and yelling out, "Just take the damn picture!"

This would be my last act as a "reality star" for a very long time.

I began to feel faint, so the twins helped me back to the couch. By this time, my entire left leg was throbbing mightily and sweat was pouring down my face and back. When I pulled up my pants leg, I could see that my knee was completely out of its socket, pushed over to the left. Queen of Denial, I was still trying to convince myself that it wasn't a big deal. *Sure, I can't exactly walk . . . and my left leg is dangling as if on a string . . . but I'll just ice it and I'll be fine.*

As the pain increased, so did my fear, and I finally caved in and let the twins call an ambulance. I began to cry (but only a little), and the first thing that came out of my mouth was: "Lucky! What's gonna happen to Lucky?" I knew nobody would be able to get inside the RV with my sidekick on the loose, and I think I cried out his name all the way to the ambulance. As much as my sons tried to reassure me that Lucky would be fine . . . I knew different.

I was still figuring that I'd just broken my leg and would be out of the hospital within a few hours with a fresh cast and a pair of sexy crutches. But at some point while I was in the emergency room, I either passed out from pain or from an injection the doctors had given me. The next thing I remember I was waking up in a fog, hearing a doctor calling my name.

"Tia, Tia . . . can you hear me? Two of us made a bet. I think you fell out of a second-story window and the other doctor thinks you got hit by a car. Which is it?"

It took me a minute to shake the fuzz from my head, and then the doctor spoke up again. "C'mon, Tia, I know you have an exciting story to tell. I watch your show."

I realized that I was in the middle of a hospital fan fest, as people around me started talking about the show. Although I knew they had only good intentions, I was feeling confused. And as I attempted to tell them the ridiculous story of tripping over a couch, I noticed it. There was no cast on my leg. There would be nothing for people to draw on and sign their get-well wishes. "What the fuck is that?" I gasped. And there it was: the scariest, ugliest contraption I had ever seen . . . a *fixator*, I was told, with four rods sticking out. I think I yelled out another profanity as the doctor gave me my dose of reality.

He explained that, in addition to my knee dislocation, my leg was basically shattered. I would have to be in this fixator, or robotic leg, for months, and then I would need to have surgery where they would insert a metal plate. The healing could take up to a year, and I would have to use a walker or wheelchair for a while, then crutches, then probably a cane.

As someone who is extremely physical and very much an outdoors person, this news was devastating—and, of course, the first thing that came to mind was my dogs. Especially Lucky, who, for all I knew, was still trapped and raging inside the RV. I needed to talk to my kids.

The second they wheeled me into my room, I called Tania, who told me that everyone had gotten together and managed to farm my dogs out to various workers and my other kids. They had managed to "double loop" Lucky so that two people could walk him together, on two leashes held taut. Not a great solution, but at least he'd be cared for. In the course of putting him back in the RV, he'd bitten Tania.

As you've already read, my kids grew up with exotic animals in their lives, so they'd all been bitten and scratched on occasion. I have some tough kids. But, as a mom, the guilt I felt about that dog bite was more painful than the oversize pins in my leg. I told Tania to leave him be, that I would deal with him. I felt as low as I have ever felt.

A minute after we'd hung up, I called for the doctor and told him that he had to sign me out or I'd just leave on my own. After a lot of fussing back and forth, the hospital had me sign some paperwork that said I was discharging myself against their judgment, and I was on my way back to apologize to both my dog and my daughter, who was trying to make light of the situation. (I'm telling you . . . my kids are so damn cool.)

It took some effort, but we managed to get Lucky and me loaded up in a minivan and over to the house where I planned to do my recovery time. I'd reluctantly agreed to stay in the city in order to be near the hospital I'd be visiting regularly, but I was reunited with Lucky, and that's what mattered.

What I hadn't counted on was the big change in Lucky. I was pretty much confined to a bed. I could barely move, and yet every time I made even the slightest attempt to adjust my position, Lucky was on top of me. He would lay his entire body over mine and begin to whine and frantically lick my face. He slept with his head over my legs (even the sore one), and I couldn't roll over without catching his watchful eye on me. This overwhelming display of affection and concern was a little unbecoming to him—but, truth be told . . . it was kinda nice.

★ ★ ★

A few months went by in this overly cozy fashion, then it was back to the hospital for what I hoped would be my final surgery. I didn't know how to tell Lucky that I would be leaving him again for a few days, but I didn't have to. He felt it. And when I hobbled out the door, the frantic barking and scratching I left behind spoke volumes. All the way to the car, I could hear him tearing things up and yelping, and when I looked back, I could see my best friend trying to peer through the crack under the door. I actually couldn't wait to get to the hospital just so I could get back to him faster. I remember thinking: *Surgery never looked so good.*

As it turned out, I was gone a week. The surgery was tougher on me than expected and the pain beyond anything I'd ever felt—and that includes childbirth. I've always had a high tolerance for pain, but this metal plate thing was brutal. And, having never had major surgery, I was stunned by the sight of all those tubes running into and out of me. The nurse told me they were for various things, including "one for pain."

As someone who has never indulged in recreational drugs or alcohol, never smoked a cigarette, and rarely even taken an aspirin, I was a little concerned about any chemicals going into my body. But the pain was beyond what even I could handle, so I thought, *Well, how much harm could a few doses of something-or-other cause?*

When I finally got back to my temporary home, life was frustrating. I was still on a walker and not very mobile.

Lucky was clingier than ever and watched every step I took, sticking close by my side as I shuffled to the bathroom and back. For the first couple of days, I seemed to be improving, but on the third morning, I woke up with uncontrollable chills. My skin was sensitive to the touch—so much so that the feel of my clothes or the sheets against my skin was painful.

I tried to sleep but was blasted awake by a very vivid nightmare, which normally I don't have. And over the next couple of days, I experienced "day terrors" that sent me into fits of unstoppable sobbing. Lucky, needless to say, was in a panic, bounding around the bed, licking me and generally freaking out.

Something was really wrong with me, and I was terrified. I couldn't control my thoughts or my body. I couldn't get a grip on my emotions.

What the hell, Tia?

I managed to call Tania, and, through my tears and hysteria, I heard her say she was heading over immediately. "Make sure you put your demon dog away," she added with characteristic snark.

When she got there, I was sitting in the living room in a chair with a blanket wrapped around me, rocking back and forth like a mental patient and moaning, "I hurt," over and over. It took Tania just a few seconds to figure out what I couldn't.

"Mom! You're going through withdrawal."

"From what?!" I screamed, as the lightbulb went off in my brain.

She asked to see my hospital paperwork, and as she looked through it, she asked, "Mom, what kind of meds did they give you?"

Because I had never taken any medications, I guess you could say I was ignorant of the effects they might have on me. "I know they gave me something for pain," I said, "but only for a few days!"

As she flipped a page in the document she was reading, Tania began to laugh. Apparently, the hospital had given me a painkiller akin to "liquid heroin," and—according to Google—nine times stronger than morphine.

I was furious. How could they have dosed me with that stuff and not even warned me about the aftermath? At the risk of sounding dramatic, I felt violated. Knowing that my body had been pumped full of chemicals made me feel dirty. But at least I knew what was wrong with me. I'd have to get clean by going cold turkey. Add to this the fact that I was confined to a small room in a house in the middle of the city. My life was starting to feel pretty damn dark.

My road to recovery was longer than I could ever have imagined it would be, both physically and emotionally. Even little things like changing my clothes were complicated and felt degrading. This level of helplessness and dependency was as strange and odious to me as the drug withdrawal, and if not for the dogs, I might not have survived it with my wits intact.

Then there was my tough-guy-turned-softie, who would try anything to lift my spirits, including tap dancing and spinning in circles until I laughed out loud. We even gained weight together, due to our mutual lack of exercise. (I wonder if Jenny Craig has a combined dog/owner program. The both of us could sure benefit from it.)

The drag on my emotions turned out to be my biggest problem. Once a self-proclaimed tough chick, I became embarrassed and fearful at the prospect of going out in public. Were those guys sizing me up to rob me because I was on crutches and made an easy target? Or were they simply getting some amusement out of watching me struggle like an idiot with a small bag of groceries? One day, after shopping, I came out to find another car parked so close to my driver's-side door that I couldn't get my unbendable leg into my own vehicle. I looked around and saw no witnesses but felt like a fool anyway, as I went around to the passenger side and painstakingly (and painfully) hauled myself in and across the seat.

Slowly, I began to venture out on walks with Lucky, who always made me feel safe and never seemed to be laughing at me. In fact, his very presence elicited respect from others. And when somebody did seem to threaten me— some picture-taking fool with no sense of compassion, for example—Lucky would quickly shut that shit down. Time after time, the no-nonsense dog with the silly ears put me before himself.

Slowly, my nightmares began to fade—perhaps due to Lucky's funny new habit. Once I was tucked up in bed, he'd crawl face-first under the covers, make one complete turn, then plop down right in the curve of my stomach, his big head tucked under my chin. Just feeling his thick muscles next to my weakened leg was enough to keep the bogeyman out of sight.

It's been three years since my unexpected and life-altering injury. As bad as it was, I have to admit that the experience taught me a few things. For example, it turns out that the old cliché isn't just a cliché at all: Our fate really can be altered in a flash. Even the toughest among us isn't invincible or indestructible, and if we think we are, we're bound to find out otherwise.

I realized that I was not as strong emotionally as I'd thought I was. Having to watch others do my errands and hold my arm so that I could get out of bed was degrading and somewhat embarrassing. And as badass as I'd always been, the pain was something I never expected and I'm pretty sure I shed too many tears to count. It sounded simple—I mean, it was just a *broken leg*—but it was a traumatic experience for me. I found myself becoming bitter and angry, just thinking about how it could so easily have been avoided. Yet here I was, "crippled."

Then there was the night I heard someone in the backyard and I couldn't even get up out of bed to save my own life. It was Lucky who jumped up to warn the intruder: "Hey, asshole . . . I can make it to the fence in 2.2 seconds . . . can you?"

These were some of the heaviest times I've ever encountered, and each day I beat myself up over that one "take it back" moment. And if one more person had said to me "Everything happens for a reason," I swear I would've throat-punched them. It felt as though I would never heal, either physically or emotionally, but that damn dog of mine just wouldn't let me give up.

Even after all this time, I still feel somewhat broken, at least in spirit. I have some pretty dreary days where I just don't want to get out of bed in the morning. Yes, even Tia Torres gets down in the damn dumps. The constant in my seemingly sinking world was a once self-centered mutt who showed this Irish lass that I was actually the center of *his* world after all. The real lesson, I guess, is that if we fight hard to survive the bad stuff life throws at us, we just might live long enough to grow up and find our purpose. That dog's purpose turned out to be me. I guess that makes me the truly lucky one, the one left holding the four-leaf clover.

8

BLUIE

The Knight in Silver Armor

Tania, my oldest "skin baby," and I have a connection that can only be described as abnormally phenomenal. As a preteen, I knew she was coming into my life, years before there was even a thought about her being conceived. I knew she would be different and odd like me, and I even knew what her name would be. Her inexplicable connection with animals was already in her DNA, and she would be my apple who would fall out of the tree and stay at the base of my trunk.

I spent most of my teen years traveling the rodeo circuit on the weekends, surrounded by horses, cows, goats, dogs, and—of course—teenage cowboys. When I wasn't competing on my horse for prizes, ribbons, money, and big shiny silver trophy belt buckles, dancing to country music in a beat-up old barn on the rodeo circuit became a huge passion.

Tanya Tucker and what she represented took over my entire being. I dressed like her, wore my hair like her, and even sang like her in the country band of which I was a part. She was confident, she spoke her mind, and was just a kick-ass chick. I felt connected to the Texas Twister and her determination to not let anyone stand in her way, so it came as no surprise that as a teenager with dreams, sitting around talking about someday finding that cowboy with whom to ride off into the sunset, I would want my daughter to be named after my favorite outlaw—though I did end up slightly changing the spelling of her name to embrace her Mexican heritage.

As it turned out, Tania's father was no cowboy and there would be no romantic ride on the back of a muscular quarter horse. There was no western sunset and no two-stepping at the local rodeo. Leaving my teenage years behind, I found myself drifting into my twenties and into a completely different outlaw lifestyle: the ghetto.

For a brief time in my life, I had become animal-less, with no four-legged companions whatsoever. During a chance stop at a downtown Los Angeles "secret" nightclub, I discovered the urban world of "lowriders": L.A. gang members and the housing projects. How I got there is an entirely different book that would take up more chapters than I could write. Let's just say that despite getting shot at and dodging more drugs than the Mexican Mafia could muster up, one good came out of the eighties (besides big hair and Madonna), and that was the birth of my oldest daughter, Tania.

Tania, her father, and I lived in the housing projects with her dad's family. I worked while my mother-in-law took care of Tania, creating a pretty sweet support system—but it wasn't destined to last for long.

Although we tried to be the typical San Fernando Valley family, it was still the eighties, and life in the ghetto was changing rapidly. There was new music, new fashion trends, and new drugs. While I was experimenting with Boy George and fingerless gloves, unbeknownst to me, Tania's dad was trying out the latest pharmaceuticals.

I had noticed a change in my husband's behavior, and his looks for that matter. He was agitated on a daily basis, and his gaunt-looking face made his weight loss obvious. But it was on a Sunday afternoon after he had been gone all night that the truth came out and hit me upside the head, both figuratively and literally.

He was "hanging with the homies" in the parking lot just outside our front door. I went out to confront him, and within seconds, the interrogation became heated. Like any typical woman, I first accused him of cheating on me. He of course denied it, but I wasn't backing down.

As his friends stood by and watched, I stepped in front of my baby daddy and gave him an ultimatum: Tell me the truth, or I was packing up Tania and we were leaving. He didn't speak a word, but his answer was loud and clear.

His fist hit my face with such force that my entire body spun completely around before slamming to the asphalt. One of his friends stepped in and held my "till-death-do-us-part" away from me as I tried to stop the world from spinning. It took me a minute or two to realize what had happened, and that's when I began to sob hysterically, then dry heave. I must've begun to mumble something like "What just happened?" because that's when one of the homeboys leaned into my ear and whispered, "It's the rocks, *chica*, it's those damn rocks." I had just been introduced to crack cocaine.

For a couple of years, I tried to hang in there. Rock cocaine came and went and made way for PCP; then heroin snuck its way around. As tough as I thought I was, I was no match for these drugs. They had taken over his life, our life, and Tania didn't know her *real* father. Ultimately, I would not have to make any decisions as to what to do about it: Crack, sherm, and smack—his three mistresses—would lead Tania's daddy down a black hole of despair. He would get shot, become paralyzed from the waist down, go to prison, and wind up dead in an alley. These events would permanently poison Tania's life and leave her with a constant sense of withdrawal from that father figure.

The years passed, and a couple of substitute fathers later, it would be a colorful, four-legged demonic-looking creature who would come to possess my daughter's world. He would also become the only true man in her life, who would offer undying and devoted companionship, and who would stand in harm's way to protect her very being, heart and soul.

It was June 2005, and I was working with animal shelters all over Los Angeles. My particular services were crucial, because in Los Angeles County at the time, individuals were forbidden from bailing Pit Bulls out of shelters. That placed the responsibility completely on rescues, and there were only a couple of us. Villalobos was busy as hell.

One morning, I received a call from a nearby facility about a Pit Bull who had been left locked up in an apartment after the tenants got evicted. They had left a note behind, saying how sorry they were that they couldn't continue to take care of "Bluie."

Over the course of my career in rescue, I've seen thousands of Pit Bulls. I've seen them in every shape, size, and color. Some of them are handsome, some formidable, some downright cute. And then there are the ones who take your breath away.

Bluie was that breath.

He was this perfect shade of the ever-so-popular gray color known as "blue" in the Pit Bull world. His white markings were perfectly placed on his muscular body, and, although I do not condone it, his ears had been cropped in what has become the traditional show design: His once natural, floppy ears had been turned into sculptured horns that made Bluie look very intimidating at first sight. But in less than a second, his "muscle butt" would begin to wiggle so hard that his tail whipped each side of his flanks, leaving welt-like markings on his metallic coat.

This dog will be a snap to find a home for, I thought. With his breathtaking good looks and seemingly obedient nature, I figured he'd be somebody's best friend in no time. And once we got him back to the rescue, we discovered that he got along with other dogs—and even cats.

It is rare that we get a dog whose original name we know, and when we do, it's usually Diablo or Felony. This dog was downright regal and deserved an awesome name of his own, but, alas, he was Bluie and he knew it. There was no changing that now. And since he'd spent his life as an apartment dog, I just didn't have the heart to put him outside in the kennel. I decided to let him stay in my house, though I told him (and myself) that this was temporary.

Tania was by now a typical teenager. Growing up in a small town, she had been raised on a ranch around many different types of animals, including wolves, throughout most of her teen years. She'd had the privilege of being a real-life Mowgli when she worked side by side with me at an exotic animal facility. She'd spent a few summer months on the back of Nellie the elephant while I trained wolves for the movie *The Jungle Book*—and throughout all of it, we ended our days taking care of our Pit Bulls. Dealing with the most controversial canines in the world had become second nature to her, as it would become for all of my kids.

So I wasn't surprised by her reaction when she came home from school one day and saw the blue dog lounging on the couch. With no dog of her own at that point, my tomboy of a daughter was like, "Whoa . . . and who are you?"

Bluie just lay sprawled out, not even lifting his head or moving one chiseled muscle in his body . . . well, except for that traditional Pit Bull tail, which went *thump, thump, thump.* I began to tell Tania the story of how Bluie's owners had left him alone in an apartment with a "please take care of him" note. But as I went on, my words seemed to drift off into the humming of the box fan battling it out with the late afternoon desert heat; Tania stroked Bluie's head and swooned over him as if he were a *Tiger Beat* cover boy.

Agua Dulce was and still is a very small town. Although you could call it middle to upper class, with its million-dollar-plus ranches, our family lived on the "wrong side of the tracks." While most in the area owned some sort of Retrievers or herding dogs, we were "those tattooed outlaws with the Pit Bulls." And at school, my kids stood out somewhat from the other students—and the brunt of the sideways looks were directed at Tania.

The teen years are always the most difficult, and like so many kids, Tania was trying to find herself. As she would describe it, she was "socially awkward," and she had only a small handful of friends. She took a stab at being a "preppy girl," but she always fell back into her comfort zone as a tomboy—a country girl. She was known as "that girl who lives with all the animals," and to the average person that might sound pretty cool, but to a group of gossipy and catty teens, it singled her out for teasing. Unbeknownst to me, the torment was much worse than I imagined.

I did begin to notice a change in my oldest. She was unhappy to the point of shutting herself away from the rest of the family, and she spent way too much time in her room.

As much as I tried to get her involved in other activities, she continued to distance herself from the family. Then I heard a rumor.

I heard that my daughter was being bullied at school—not just in the usual way, by girls making fun of her, but by a group of boys from the football team. Maybe because Tania dressed like a typical country girl/tomboy, they didn't just make fun of her but also tried to pressure her into doing things she didn't want to do. I heard how these boys would corner her in the hallway and basically tell her that if she did "this and that," she would be one of the cool kids. They laughed at her and publicly shamed her.

It took all the strength I had to restrain myself from jumping into my truck and heading out into the desert to hunt them down. Instead, out of nothing but major respect for Tania, I sat her down and told her what I had heard.

Tears were shed and I felt horrible that she had felt the need to keep it all from me for so long. With my typical outlaw attitude, I offered to go out and tear some heads off—but Tania, being a peacemaker, begged me to stay put. She explained that it would only make things worse and that she wanted to figure out a way to handle it herself. Although I wasn't thrilled about her optimism, I knew it was important to empower her with the strength and confidence she needed.

Then an idea came to mind.

I wanted my daughter to feel safe when she was out and about without bringing attention to the situation, embarrassing her, or putting her in a worse position. I wanted Tania to go from feeling like a victim to embracing her strong, fiery Latina roots. And I knew of a boy who would watch out for her and could back up his shit if shit went down.

With his eighty pounds of ready-to-launch muscle, Bluie was my pick of suitors for the child I called "my apple." Like her mother, Tania was reserved and somewhat reclusive, yet we both possessed that "quiet storm" within. Having Duke at my side for all those years had not only empowered me but had given me a sense of security beyond measure. I wanted this for Tania, and I knew the blue dog had what it took. I wanted Tania to understand that this was a "man" on whom she could depend, one who would never make fun of her and would love her unconditionally. It was then that Bluie moved into my lost child's room.

It didn't take long before Tania and Bluie clicked, as she would say, "like peas and carrots." He followed her constantly throughout the house, into the bathroom, to watch TV, to take a shower, and pretty much never left her side. But it was on one trip to the local McDonald's that Bluie "marked his territory," and the relationship made an even bigger impact on their lives.

Tania and her two closest friends decided to drive down to the fast food joint, which was pretty much the only place to go and hang out. Now that Bluie had become a constant in my girl's daily routine, there was no question that he would accompany the three friends to lunch. He got along well with any humans and pets who crossed his path, all the while being very protective of Tania, just as I'd hoped. You could see it in his stare, which reminded me of that famous James Dean photo—the one where he's leaning against an old black roadster motorcycle, cigarette dangling, his eyes saying, "I dare you."

The McDonald's parking lot was scattered with groups of teenagers, some sitting on the hoods of cars and others inside their vehicles, taking cover from the hot desert sun. Various genres of music blared from car stereos as boys with shredded jeans and band T-shirts skateboarded to the beats. Tania and her friends opted for the comfort of their car. Bluie had just settled down in the backseat for a nap when a screech of tires announced the arrival of a new group of boys—the football players who'd been menacing my daughter.

Conversation inside the car stopped. Bluie sat up, his "horns" quivering at the sense of anxiety pouring off of his best friend. As the boys got out and walked past the car, Tania slumped a little in her seat and attempted to shield her face with her hand, but one of the gang spotted her and sauntered up to her window, and the shit-talking began.

With nowhere to go, Tania did her best to ignore him and quietly grasped the crank to roll up her window—but not fast enough. The boy reached for Tania's arm, and it was as if a nuclear bomb had gone off. Bluie, who had been patiently lying low in the backseat, exploded and, like a missile, catapulted himself over the seat and halfway out the window. His barking and roaring easily drowned out the opening chords of "Bad to the Bone," which had just come on the dashboard radio, providing an apt soundtrack for the occasion.

As the boy and his buddies jumped backward, the entire parking lot seemed to freeze for a few seconds. Then it came—an outburst of spontaneous laughter from inside and outside the car. The vision of that big ol' block-headed Pit Bull hanging halfway out the window, barking as if to mock the now-running-for-their-lives former tough guys, left an ear-to-ear grin on my daughter's face. Never again would she have any problems at school. The new "bully" in her life had just put things into perspective, and Tania suddenly felt empowered.

From that day on, Bluie and Tania were inseparable. Things did get better for her socially, but her unquenchably rebellious nature caused some conflict with her teachers, so in her last year of high school, I homeschooled her. Bluie, of course, could not have been happier about that, and their bond strengthened to the point where I could no longer go into her bedroom without begging for mercy—nobody could. In fact, anyone who wanted to enter Tania's airspace pretty much had to ask permission from that blue dog.

It was kind of a family joke, really. If Tania decided to sleep in, I couldn't get near her to yank the covers off. If she played her music too loud, I could barely crack the bedroom door open to complain before Mr. Buns of Steel would leap at it and slam it against my forehead. Bluie was in charge of Tania now, and I was secretly quite relieved about that.

Tania qualified for her high school diploma, then began to work for me full time. She moved from taking care of the wolves to working with the dogs, all the while becoming an adept caregiver for our more exotic friends: tigers, lynx, cougar, leopard, and bears. Bluie had managed to do what I couldn't—or, at least, not on my own. He'd given her the courage and confidence to face the world head on.

★ ★ ★

No one could've predicted the monster that my "mini me" and her faithful companion would soon encounter. Bluie would be forced to assume the most dangerous role of his life when Tania entered into a relationship with a young man who had come to work for me.

Although a little rough around the edges, he was polite and seemed to want to make something of himself. He had a great work ethic and was always very respectful. Tania ended up being with this guy, who I will call "Dick," on and off for years. I took it as a typical young romance with the normal ups and downs, and I really didn't think much about it. Unbeknownst to me, there was an unthinkable storm brewing behind closed doors. It would take Bluie's increasingly aggressive behavior to clue me in that something was going on.

Bluie had always been protective over Tania, but now his protectiveness went from growling and barking to actually running up on a person and standing his ground. When he pressed his muzzle against your leg, you could feel the low rumble from deep within him resonating through your entire body. In those moments, you dared not move any closer to Tania for fear of losing a limb.

It troubled me to watch this behavior get worse, as Pit Bulls are very sweet with people by nature. Something was very wrong, and Bluie was trying to tell me a story. I just needed to open the book and read, even if it was between the lines.

After much probing and interrogating, Tania finally poured her heart out to me, confessing that she was ashamed of herself and angry at the same time. Dick had been abusing her, both physically and emotionally. It had started with a shove against the wall, which had driven Bluie into Tasmanian Devil mode. Frothing at the mouth, barking and snapping, he'd positioned himself between Tania and her aggressor until Dick backed off. But, like so many women in abusive relationships—including her own mother—Tania had wanted to give the guy another chance.

Despite my attempts to get Tania to end the relationship, the abuse went on for years, and each time Dick lifted a hand to Tania, it was Bluie to the rescue. The heel learned to bide his time, to wait for one of the rare occasions when Bluie wasn't around, before he'd scream at Tania or toss her against a wall like a rag doll. This only made Bluie more watchful. Like a panther on the prowl, he'd appear from out of nowhere and put his own safety at risk to protect his girl.

Sitting with me on the couch, shedding buckets of tears, Tania told me story after story of how she would pretend to be asleep whenever she heard her boyfriend stagger in drunk. Bluie would position himself across Tania's body, growling from the depths of his soul and waiting for the evil creature to come slithering in the door. Tania would peer out from under Bluie into the darkness, hoping Dick would finally stumble back out to the couch and pass out—which he usually did.

As I listened to my Doppelganger of a daughter tell me grim tales of mental and physical abuse, my heart broke into a million pieces. As hard as I had worked to raise her to be strong and avoid the mistakes I'd made, I'd clearly failed miserably. And to realize that, if not for her brave and faithful dog, it all would have turned out much worse . . . well, I shuddered to think how bad it might've had been. I thought about my own past and what she had witnessed as a child. The guilt set in, and I hated myself.

Then I looked over at Bluie. As tears streamed down Tania's face, I swear his eyes filled with the same. You could feel the concern overcome him as he licked the salty water pouring from my daughter's eyes. The love this dog had for Tania was undeniable, and I thanked him for saving her life when I couldn't be there. He had made many sacrifices in putting her life before his, and he had made them unconditionally.

Bluie and Tania vowed to never let anyone come between them again. Even at her young age, Tania had endured something that no woman should ever have to go through. She told me that Bluie was now her main man and that, unlike all the other guys who had failed her, including her own father and her first serious relationship, her true love story stood right before her and they had made a pact: till death do they part.

★ ★ ★

The decision to move our entire family from California to Louisiana took us on one of our wildest adventures. We had to make about eight trips back and forth across the country to move dogs, cats, household goods, kennel supplies, vehicles, and everything else. Many of those trips were made in a donated old school bus, and one of them would end up putting Tania and Bluie in a situation worthy of an episode of *I Love Lucy.*

It was August, and traveling the I-10 interstate through the deserts of California, Arizona, New Mexico, and Texas was brutal. Tania, Kanani, and Jake (one of our first parolees from California) were making trip number three, carrying some household goods, Tania's parrot, and their own dogs: Bluie (Tania's), Snorky (Jake's), and Jean and Rogue (Kanani's). The school bus was kind of an ingenius contraption—a mousetrap on wheels—complete with a homemade air-conditioning system devised by the twins out of ducts and tubing and all sorts of nuts and bolts. The interior looked laughable, but once the thing was in motion, it functioned like a charm, thanks to a home generator secured to the outside rear.

The trip was going fairly well as they crossed over the Texas border . . . and then the generator died. Despite Kanani's best efforts as a mechanic, they just couldn't get it up and running again. This meant they had one miserable trip ahead of them, through the hottest part of the country. As the bus filled with dry desert air from the now open windows, several of the animals started to pant, but not Bluie. Despite the overwhelming heat and Tania's attempts to send him to the back of the bus, where he could sit in front of the battery-operated fans they'd rigged up, he wouldn't leave her side. He just sat quietly stoic and leaned his shoulder against her very sweaty leg.

We had already made a couple of trips to our new home state, and in our previous passes, we'd encountered no trouble with the border patrol in Texas. One or more of the officers usually recognized the kids from our TV show and launched into a conversation about dogs. But on this cursed day, the kids' luck—or fame—had seemingly run out. The officer on duty insisted that his drug-detecting dog "smelled something." I guess it didn't occur to him that even a dog trained to do police work is still . . . a dog. Between the pungent smell of the parrot and the various dogs that had been on the bus, it was no wonder this dog got excited.

There was going to be a search.

So there they stood on the edge of the interstate, Kanani wearing only board shorts and flip-flops, holding on to his barely leash-trained Dingoe. Jake was in boxer shorts, hanging on to Snorky, and Tania was in a sports bra, parrot in one hand and Bluie's leash in the other, as he barked his block head off. The stars of Animal Planet's hit show *Pit Bulls & Parolees* looked more like the cast of *Trailer Park Boys*. It was kind of a nightmare at the time, but the rest of us got a good laugh about it later, at their expense. Needless to say, the bus was clean—of drugs, anyway—and they managed to haul their sweaty asses the rest of the way to New Orleans without major mishap.

★ ★ ★

The move to Louisiana gave us all a new start, and we were happy to be looking at California, and our various troubles there, in the rearview. All four of my kids found new significant others, and life was good in the relationship department. Tania even found a man who loved Bluie as much as she did, and Bluie immediately accepted him. This was a good sign. On the day that Perry proposed to her, Tania sat down with Bluie and had a long talk with him, reassuring her first and best man that she needed him as much as ever and that he would never be replaced by a mere human.

The new family merged perfectly. From the get-go, Bluie not only accepted Perry but adored him. Once Tania saw that Bluie was giving his paw of approval, it was okay to set the date—though she still felt a little like she was cheating on her man of steel. She and Perry decided on Halloween 2013, and they made sure to include Bluie in all of the plans.

Ceremonial plans went into high gear, and one thing that had to be taken care of was the engagement photos. A good friend of ours who also happens to be an awesome photographer came out from California and took the most breathtaking shots of Tania and Bluie as a couple in love, and then of the new human couple flanked by the gorgeous Pit Bull. The pictorial memories of the Halloween wedding itself were even more spectacular. Decked out in black and red, the bride, groom, and faithful companion looked like they had walked out of an old history book. Despite the fact that he was giving away his girl, Bluie looked happy and content. It was almost as if he was preparing for something.

The new family moved into their dream house in the Upper Ninth Ward area of New Orleans. Because the house was built over a basement (unusual in New Orleans), it stood tall above all the other homes in the area. At a hundred and fifty years old, it looked like a place where the Addams Family might live. Bluie was able to perch himself on its high porch like a gargoyle protecting his territory. As the Pit Bull protector scanned the street looking for potential enemies, he'd take a second to turn his head back to gaze through the window.

Once his princess, Tania was growing up and becoming something much more. And yes, Bluie did feel a little bit jealous that there was to be a new king in the castle—but he was happy for his girl. He loved that she laughed so much more than she had in those days back in California. She was more self-confident, and she had dreams and goals. This was all he had ever wanted for her. Bluie must have been conflicted, knowing it was time to allow Tania to live her life and prepare for a great future. He had raised her from a confused and lost teenage girl and kept her safe from life's monsters. But the time had come to allow the love of his life to create her own kingdom.

A few months later, Bluie suddenly fell ill. Tania took him

in for what she thought would be a routine vet visit, figuring she'd get him some antibiotics and he'd be back to his normal blue dog ways. Nothing in the world could've prepared Tania for hearing that her best friend had cancer—an aggressive form that couldn't be fought.

Our family pack circled the wagons and came together for emotional support. We understood the pain that Tania was experiencing was unbearable. She said that she couldn't go on with her life without Bluie. But somehow, she managed to pull herself together and give him the dignified ending that he so deserved. He would not want pity. He would not want tears. He was a strong dog who would want everyone around him to feed off his strength. Tania did her best to honor this final wish.

As a family, we began to prepare for Bluie's final chapter. Tania had always said that Bluie was her once-upon-a-time, her fairy tale with a happily-ever-after, but his story was coming to an end, and we knew that his remaining time on Earth would not be long. We had to create one final and lasting memory—a story to be told for years to come. It had to involve something that Bluie loved to do. Tania decided that she, Perry, Bluie, and his scruffy little pooch of a sidekick LuLu would go on one last road trip together.

They loaded up the camper. Despite the fact that Bluie didn't feel all that well, he couldn't wait to jump up and find his place up between the front seats to watch out the picturesque windshield. As they headed out of the Crescent City for a destination unknown, Tania looked over at the still regal dog. This would be their final sunset together, their last night sleeping under the stars.

Tania popped in her and Bluie's favorite traveling music—"Free," by the Zac Brown Band—into the dashboard stereo: *Just as free / Free as we'll ever be . . . / Me and you.* It was "their" song.

Within a couple of days of returning home, Bluie told Tania that it was time for her to be free of him, to live her life as a strong and passionate woman. He was ready to move on and hand her over to the new man in her life. Drawing from his courage, she finally made the brave yet agonizing decision to free him from his pain.

Tania's knight in shining armor had slain enough dragons in their lifetime together to make her feel safe forever. As a princess who had grown to become a queen, she imagined her prince charging off to save yet another damsel in distress. Maybe, just maybe, in a land somewhere in another time, there is another little princess who will have her happily-ever-after because of the silvery dog with the colorful name.

9

TAZ

The Little Dog with Nine Lives

My very serious leg injury, followed by two surgeries and seven months of strict bed rest, altered my life forever in ways I never could've imagined.

Honestly? For me, the emotional torment was the worst part. I'd been extremely active and independent all my life, without really giving it a second thought. Overnight, I went from a state of total mastery of my world to barely being able to move without a helper, a walker, crutches . . . and a lot of pain. I can only describe it as *degrading*.

Simple things like taking a bath and getting dressed took hours. As the weeks passed, I kept trying to convince myself that things were getting better; that my hair couldn't *possibly* be falling out as a result of all that anesthesia from the back-to-back surgeries; that my ongoing nausea from the pain meds I'd taken was all in my head and not my aching stomach. If only the medical professionals who had become the newest fixtures in my life had taken the time to explain what their "miracle drugs" were doing to me, I might've at least understood what was going on. The reality is that my previously chemical-free body had now been violated, and I felt "dirty." As a result of it all, I was confused and growing more depressed by the day.

As I write this, it's been a couple of years since that couch attacked me, and I still walk with a limp, can't bend my knee, and can't shed the weight I've gained from being so sedentary. Add to all of the above a general loss of energy that makes me feel like someone twice my age—someone who will never have the opportunity to accomplish everything in life I want to do and to make up for lost or wasted time along the way.

Rereading the above, it sounds kind of whiny, and that is *not* the Tia I have ever been or want to be. I've never had any patience for people who feel sorry for themselves or run to the doctor for every little thing. Before the accident, the only times I'd seen the inside of a hospital were when I gave birth. (I did once break my foot when I tripped over the twins' pet rooster—but I just wrapped it up tight with an Ace bandage and let it heal on its own.) I remain determined to snap back to my "real" self; I just hope that I have the patience and resilience to get there.

In trying to recover from something so big, I couldn't help but think about all the dogs who have come into my life having endured so much more. I've met more victims of dogfighting than I can count who lay on the vet's exam table, torn to pieces yet wagging their muscled tails and licking any face within reach, simply out of gratitude.

At Villalobos, dogs have come through our doors having been shot, stabbed, set on fire, hit by cars, with broken limbs dangling . . . yet they all—every single one of these brave souls—accepted their pain like champions and never complained. In fact, quite the opposite: They maintained upbeat attitudes, wiggly butts, and a sense of forgiveness toward the beings we call "human" who had caused their suffering in the first place.

But, of all the hundreds of dogs who have come to us under dark circumstances, one stands out for having endured more trauma, pain, and torture than any other.

Named after that cartoon tyrant the Tasmanian Devil, Taz made such an impression on my life that my thoughts still turn to him the second I feel myself lapsing into self-pity. Taz wouldn't tolerate that for a second, and I know it.

On August 29, 2005, Hurricane Katrina blasted the Gulf Coast like a pissed-off bitch on a mission. Labeled Category 5 at her worst, she bore down on Louisiana with such force that the unthinkable happened: Several levees were breached, causing massive flooding throughout the southern tip of "the boot."

My family and I watched from the comfort of our desert home in southern California as the TV news showed thousands, both human and animal, fighting to stay alive. I became inconsolable, and my emotions got the best of me.

I wanted to do something but felt so helpless. Even if I'd

had the means to simply pack up and go, I had a large rescue to run, not to mention human children as well. Between the TV news and the Internet, the stories were horrifying. Personal friends of mine who'd gone to help sent me full accounts of what they were going through in assisting with the animal rescue efforts.

One of my closest friends called me from New Orleans and told me that he actually regretted going because of what he was witnessing. He said that he was starting to have nightmares, the result of an almost impossible time trying to cope with what his eyes saw and his nose smelled.

In spite of this, I could barely sit still for wanting to go down there. I heard rumors that there were more Pit Bulls running loose and swimming for safety than any other type of dog. An estimated five thousand of them were being housed at a large evacuation site just outside the city. And that's *just* Pit Bulls! There were many thousands more animals—dogs, cats, and livestock—besides. There was just no way to imagine this kind of hell on Earth, and I struggled to get my mind around it as I went about my business two thousand miles away.

The weeks dragged on as I followed the rescue efforts from afar. Then, in October, I got the call that I'd been expecting but dreading. Many of the larger animal rescue groups that had traveled to the Gulf after the storm were packing up and heading home, leaving hundreds of displaced animals behind. Even the local rescue workers were exhausted and had to get back to their lives, jobs, and homes. They scrambled to find places for the last remaining animals.

A group of nearly fifty Pit Bulls dubbed "the bottom of the barrel"—those whom no one had seen fit to save—were at an evacuation location known as the Lamar Dixon fairgrounds. These dogs had health or behavioral issues that steered people away from taking them. These were the true underdogs. And Villalobos was being asked to take them.

Along with offering some donations of money and supplies, we agreed to take the "last ones standing," and, just like that—within twenty-four hours—they arrived at LAX via cargo plane, met by an overwhelming crowd of dog lovers, media, and a soon to be very tired VRC staff. The dogs were stacked in crates and cages. Some seemed thrilled to be in a new place while some were so traumatized that they crouched, shaking and trembling, at the back of their enclosures. By the time we'd gotten them all off the plane and into a large trailer, the day had turned into a cold night. The newly transplanted dogs quivered in their crates, some from excitement and others nervous about the unknown that awaited them at their newly formed M.A.S.H. unit back at Villalobos, dubbed Camp Pitrina.

The weeks that followed were brutal. As we tried to get the displaced pups settled and acclimated, we frantically worked to construct comfortable facilities for them in our front yard. Meanwhile, back in Louisiana, more desperate dogs awaited help, and their stories came flooding into our inbox on a daily basis. There was a limit to what we could handle, but the saga of one particular dog stood out among the hundreds.

One of our longest-serving volunteers, a dedicated woman (and fellow Pit Bull owner) named Becky, had flown down to Louisiana immediately after Katrina to help out with the animal rescue. Her particular skills had come in very handy, as many of the volunteers there were not skilled at coping with particular behaviors of Pit Bulls, including their sudden bursts of energy and strength. These well-meaning people were simply overwhelmed by taking care of the dogs and were searching desperately for experienced Pit Bull rescue groups to take them.

Becky had hooked up with a group out of Washington State that had no Pit Bull experience at all, and they were becoming increasingly desperate to unload the sixty-one dogs they had accumulated at their rescue camp site in a local Winn-Dixie grocery store parking lot. In a phone call, Becky told me that she was concerned about the fate of the dogs. All of the larger animal organizations had already packed up and returned home, leaving the smaller grassroots groups to fend for themselves with few resources. No one knew what to do with the remaining Pit Bulls.

Then, out of nowhere, a guardian angel appeared—or so it seemed. She told the Washington group that she was a Pit Bull rescuer based in Arkansas and that she could take all sixty-one of the dogs with her. But something just didn't sit right with Becky. Meanwhile, at that point, she had become very attached to one particular dog. He was small and compact, with pointy little ears that complemented his devilish attitude. Becky had named him Taz and she decided that she didn't feel right about shoving him—or any of the dogs, really—into this woman's gooseneck horse trailer.

But the leader of the Washington-based group was all too willing to load every single crate into the rescuer's trailer without so much as verifying who she was. She'd billed herself as a "rescuer," but, unfortunately, throwing that word around in the animal rescue community was becoming increasingly deceptive.

Feeling as if she were in a scene from a movie, Becky stood beside the road as the Louisiana sun dipped below the Mississippi River and watched as her little friend Taz drove off to what she hoped would be his bright new future. Soon after that, Becky returned home to Sacramento, but as the days passed, her fears threatened to overcome that hope.

Once she'd gotten home, she called me and we had a long conversation, in which she relived her experiences in New Orleans. She described the trip as by far one of the worst and best things she had ever done. It had felt good to be able to help the dogs, but she'd seen so much sadness that it had changed her forever—and on top of all of it, she just couldn't get the image of that little white Pit Bull with the brown eye patch out of her mind. She wondered how he was doing and if he would ever find his owners.

Early on in the aftermath of the storm, there were some rescue procedures put into place by the bigger agencies. Anyone who took in displaced Katrina dogs was mandated to keep them for a specified period of time, in order to give

their owners a chance to find them. They were all photo-
graphed, and their pictures and particulars were posted on
the Petfinders website. As rescuers, we were instructed to
monitor the responses and act on any calls or emails we got
that seemed legitimate. I have to admit, out of all of the
chaos, a certain kind of order did prevail.

All the rescue groups who'd pitched in were given a
"dowry" of sorts: $1,000 per dog, to help offset the costs
of their care. And anyone who succeeded in reuniting a dog
with its owner received a $500 bonus to help cover its travel
expenses back to Louisiana. This seemed like a generous
incentive to me—but of course, people being what they are,
there were those who were busy figuring out ways to game
the system for their own benefit.

As the weeks went by, a variety of dog owners came forward to claim "their" pets, and many dogs were surrendered
with no questions asked. Some of the more unethical rescue groups were putting dogs on planes and shipping them
to anyone who claimed ownership—including those who
might be involved in dogfighting, drug dealers, who knows
what. It was a quick and easy way to claim their bounties.
Just as we were beginning to figure this out, Becky got a
phone call that ripped her heart out.

It turned out that some other legit rescue people who had
been in New Orleans with Becky had had concerns about
the Arkansas Pit Bull rescuer as well. One of them had sent
some people down to check on her facility, and what they
found was acres and acres of hell on Earth. Approximately
five hundred dogs, still zip-tied in their cages and crates,

had been dumped in the middle of a field. Many of them lay dead of starvation, dehydration, and heat stroke. There were mountains of trash, including piles of black garbage bags stuffed with more canine corpses, and even more lifeless bodies scattered and rotting around the property.

The deceitful woman who'd billed herself as a Pit Bull rescuer had simply cashed in on the suffering of the displaced Katrina dogs and dumped them to endure a slow and agonizing torture worse than the storm itself.

The hunch that Becky had carried with her had been right all along. Now the big question weighed heavy on her heart: Was Taz still alive?

By the time law enforcement arrived at the scene of the crime, a handful of reputable dog rescuers and Pit Bull enthusiasts were there, along with news crews. Animal lovers from around the country descended to witness what could only be described as a "canine apocalypse." They walked slowly up and down the rows and rows of cages, searching for signs of life and, sadly . . . death.

As Becky's "boots on the ground" contact passed each cage, she hesitantly looked in, both with hope and fear of finding Taz. The intermittent sight of small white paws sticking out of a black garbage bag nearly undid her. Tears streaming down her face, the volunteer slowly pulled back the black plastic, relieved to see that it was not Taz but, at the same time, feeling guilty that this was her reaction at all.

Emotionally exhausted, the volunteer working on Becky's behalf peered down into one more cage and—there he was. On his side at the bottom of the same wire cage in which

Becky had placed him a couple of weeks prior was Taz—alive, but barely. He was skin and bones and unable to lift his head when she said his name, but she knew him immediately from his description. Having endured the worst natural disaster in his hometown's history, he had now gone on to endure pure human evil.

★ ★ ★

For the second time, the "little dog that could" was rescued, but this time he was brought to Villalobos Rescue Center, a place where he would be safe from all that could harm him.

Taz thrived in our care and, as we'd predicted, became quite the little pistol. His perfect brown eye patch and cocked head gave him the look of a living, breathing cartoon character, and we knew he'd make a great addition to someone's family—but every now and then, his quirky behavior would take a turn that might not suit some people. So, as cute and irresistible as he was, we decided that once his mandated "stray hold" was up, we would look for that special adopter with the patience and understanding Taz might need. One thing was certain: He deserved his perfect happy ending, having been to hell and back—twice.

In keeping with the twists and turns that seemed to characterize Taz's short life, we received a phone call one morning from a man swearing up and down that he was Taz's owner from New Orleans. At first we were like "Wow! Great news!" But this quickly became "Crap . . . we don't want to let him go." The little guy had been through so much. We weren't really sure we trusted anyone else with him—even his original owner. But we'd promised to uphold the rules, so, after doing our due diligence and determining that the man was indeed Taz's true owner, we made his travel plans.

Another rescue group might have simply put Taz on a plane, but we opted to drive him down to the Crescent City ourselves. He'd traveled the highway to hell once before, and we were determined to deliver him safely and see for ourselves where our little friend would end up this time. I didn't have the time to make the trip myself, so I recruited two of my best volunteers to take the cross-country trek with Taz from our high desert ranch back to Louisiana.

On the designated morning, we were all feeling sad and anxious. We procrastinated by making a big production out of loading all of Taz's favorite toys and blankets into the van. Of course, we hoped that his owner would note and appreciate the loving care we'd showered on his dog, and that he'd be patient with any symptoms of PTSD the poor little soul had developed. So many people from around the country had been involved in saving Taz, and he would carry a piece of our hearts with him to his new/old home.

But . . . something about Taz's owner just didn't seem quite right to me. I just couldn't put my finger on the problem. I mean, his owner seemed excited at the prospect of reuniting with him and all, but there was one phrase, one sentiment, he kept repeating that really bothered me:

He's my dog and I just want him back.

To my ears, it sounded as if this guy was more interested

in claiming his property than reuniting with his beloved companion. But what could I do about it? There was no doubt that Taz was his; he'd described him to a tee, right down to the scar on his wiggly little tail. So, against my instincts but in deference to the rules, we tearfully watched as the little ring-eyed dog with the crooked tail gazed at us through the back window of the van, the desert dust swallowing him up as he bid farewell to California.

A few days passed and we grew eager to hear back from our volunteers about the happy reunion of man and beast. When the call finally came in, we gathered around, wanting desperately to experience a sense of relief that at least one of our Katrina dogs had found his way back to a loving family. Nothing could've prepared us for what we heard on the other end of the phone.

Sandy's voice was so full of panic that I had to calm her down and tell her to talk more slowly so that we could make sense of what she was saying. She took a deep breath and started her story over.

When they pulled up to the house, the man we'd talked to wasn't home. His wife told them to go out back and hook Taz up to the chain on the fence.

Hmmm.

Giving her the benefit of the doubt, Sandy thought, *Maybe their property has been so severely damaged that they're making do with what they have.* Shaking her head, she held Taz's leash and walked behind the house—where she came upon what she described to us as a Pit Bull house of horrors.

There, in the residual mud and muck of Katrina, were

several emaciated Pit Bulls chained to a fence. Nearby, in a dilapidated cage, were a mother dog and a litter of puppies, lying in layers of waste and feces. Next, she saw a skeletal tan Pit mix whose tow chain had kinked up so badly that he was hunched over from the weight of its enormous links.

Sandy had seen quite enough. Her heart racing, she kept a tight hold on Taz as she backed out of the yard.

"Where do you think you're going?" the proprietor of this hellhole called out. Thinking quickly, Sandy responded with a fabricated story: "Oh . . . I forgot some of the dog's toys . . . I'm just going back to get them . . ."

The second she reached the rental car, she called me.

I knew I had to make a quick decision that could jeopardize the lives of my volunteers. As we'd been talking, more people had emerged from the house in New Orleans. They'd sensed what was up and were beginning to surround the car. Clearly, they wanted their "property" back.

Setting aside the rules we'd been told to follow, I yelled, "Get the hell out of there!" into the phone and, just like that, Sandy motioned for her accomplice to get into the driver's seat. In one swift move, she picked up Taz, jumped into the car like Robin into the Batmobile, and off they sped, stirring up a storm of gravel as they accelerated.

Forty-eight hours later, Taz was back in California, having escaped yet another terrible fate. We did get a few threatening calls from the owners from hell, which we took great pleasure in ignoring. He was *our* dog now, and not even a Category 5 could tear him away from us.

As traumatic and dangerous as this failed reunification

had been, Taz came out of it a hero. We were able to contact animal control and they confiscated all the dogs out of that yard. They eventually made their way out to California, where they joined the Villalobos family and were eventually all adopted out.

It seemed ironic that five years later, when we picked up and moved to Louisiana in 2011, Taz was once again making the trek to New Orleans—this time as a permanent VRC dog. We'd gotten his PTSD under control, but he'd never become adoptable, and that was fine with us. Who could blame the little guy for having a few personality problems after all he had gone through?

Taz was one of a handful of Katrina dogs we still had with us. Relocating them back to their home state seemed both surreal and serendipitous. They were among the triumphant survivors pouring back into the city they'd loved and lost, and in that regard, they were conquering heroes. At the same time, New Orleans was still shrouded in some sadness and hadn't entirely become her old self again. Did the dogs have any memories of what had happened there? Would "coming home" be too traumatic and haunt them? We were conflicted.

True to his feisty Tasmanian Devil ways, our little spitfire thrived. He trotted around the new VRC site like he was the baddest little dude on the block—as if he knew he was on his home turf. But this time, he was a member of a big, loving gang who would never let harm come to him again.

Because of his amazing story, it was decided that Taz would be the new face of VRC New Orleans. Reluctantly, I gave the okay to retire my dear departed Duke from our business cards and letterhead and swap in an image of Taz's pint-size bad self in his place. Having survived more drama, trauma, and near-death than any dog in Villalobos history, Taz was the perfect symbol of all that we believed Pit Bulls stood for: strength, resilience, and spirit.

As tourists began to shuffle through our new facility, they'd hear Taz's story and a look of pity would cross their faces. But the last thing Taz wanted was pity. He'd show off his powerful whip of a tail and shrill voice even as the tour guide was going on about his battles with Katrina and unprincipled rescuers.

Taz never considered himself a victim and never thought he'd meet an enemy he couldn't beat. Neither did we, I suppose—but one nemesis proved tougher than even Taz. After overcoming nature and neglect and racking up more cross-country miles than a long-haul trucker, our little four-legged typhoon succumbed to cancer on November 10, 2013.

As we sat around our warehouse in the Upper Ninth Ward and reminisced about his roller coaster of a life journey, copious tears were shed and memories were cherished. We laughed about his spiky attitude when he didn't get his way and his flair for throwing a stinkeye . . . but mainly, we talked about his heart. Most animals—and humans, for that matter—would have given up in the face of all that Taz went through, but not him. He always fought back with every ounce of strength his little muscle butt could muster up.

Later, sifting through the photos that documented Taz's crazy life, somebody repeated an old cliché that had never struck us as truer: "It isn't the size of the dog in the fight that matters, but the size of the fight in the dog." I like to think that, wherever he is now, Taz is shaking it all off as if it were nothing, cocking his cute little head, and saying to all those who had hurt or neglected him: *Bite me.*

Take a chunk out of each one of them for me, little dude.

ABOUT VILLALOBOS

Villalobos Rescue Center has become so much more than just the largest Pit Bull rescue organization in the world. We've been through more battles than I care to count and despite being social outcasts, we remain proud and true to what we stand for: the underdogs.

From dogs to cats to livestock and other creatures, we've managed to embrace every member of the animal kingdom. And then when you throw in the human aspect, well, the empire just continues to grow.

But I think the biggest secret that we hold is that despite how successful our TV world is, beyond the cameras . . . the struggle is real. Most of the time, too real. Without the fans, the viewers, the supporters, the volunteers, and all the other people who help keep our doors open, we couldn't have made it this far. All we ask is that you never forget about us, or them . . . yes, I mean the dogs. Every now and then, and after watching the newest episode, just remember that we are real people, and these are real dogs.

★ ★ ★

Remember to check us out at www.vrcpitbull.com, or our Facebook at https://www.facebook.com/VillalobosRescue Center.

The cameras only give you a minor glimpse into our lives. Join us daily and stand beside us as we continue to fight for those who can't defend themselves. We are The Underdawgz.

ACKNOWLEDGMENTS

I think the first to be thanked are also the inspiration for this book: the dogs. Without them, there would be no Villalobos Rescue Center or *Pit Bulls & Parolees*. But more important, the purpose for me being on this Earth would have less meaning. The dogs have not only taught me, but my kids as well, the true meaning of being human. For this, I cannot thank them enough. There are just not enough words.

And then there are the people of Villalobos Rescue Center: the workers, volunteers, and, yes, even those kids of mine . . . all eight of them. Without each and every person involved in running this dog rescue, there is no way that we would've survived this long. The number of smiles and tears put into each day cannot be measured.

And what about those fans, eh? We have the most dedicated and loyal ones ever. The amount of support they've given us throughout the years speaks volumes, and I don't think they realize how important they really are. The fans and supporters are the blood that runs through our veins. They are our lifeline.

Then there's "The Planet," my term of endearment for our network, Animal Planet. Talk about a group of people who took a leap of faith on probably one of the most controversial topics on television today. It's been a long and bumpy ride, but they've stuck it out and kept us from driving over the cliff. I'm so glad all those *other* networks turned us down.

Never in my wildest dreams would I have imagined that prison would bring us together with the most awesome production company—44 Blue Productions. If not for the shared passion of helping those "beyond the walls," our paths would never have crossed. You have become a second family to us, and having you by our side each season gives us that sense of security that only a true family can offer.

There is only one way to thank my publisher, editor, and agents, and that's to say, "I'm so sorry that I drove you all nuts. Didn't anyone warn you about me?" What can I say? I'm a passionate person when it comes to my writing, and to be given the opportunity to share some of my life with everyone, well . . . it had to be just perfect. Thank you all for putting up with me.

My final thank-you goes to those who inspired Villalobos Rescue Center from the start. To the wolves and wolf hybrids, past and present. It was your spirit, your family values, and your loyalty to me and my family that embodied the true meaning of "the pack." You gave us strength when we were weak, laughter when we were sad, and light when it was dark. Here's to many more full moons.